# EMPLOYING COMMERCIAL SATELLITE COMMUNICATIONS

## WIDEBAND INVESTMENT OPTIONS FOR THE DEPARTMENT OF DEFENSE

Tim Bonds
Michael G. Mattock
Thomas Hamilton
Carl Rhodes
Michael Scheiern
Philip M. Feldman
David R. Frelinger
Robert Uy
with
Keith Henry
Benjamin Zycher
Tim Lai
David Persselin

Project AIR FORCE
RAND

Prepared for the
United States Air Force

The Department of Defense (DoD) is considering increasing the amount of communications it owns and leases. Decisions on how much communications capacity to obtain, and how much will come from DoD-owned assets, will affect Air Force investments in new communications satellites. This report assesses military use of commercial wideband satellites by evaluating their effectiveness across several characteristics defined by the United States Space Command. The cost of buying or leasing commercial systems is then found and compared with the cost of buying military systems with commercial characteristics.

The study was undertaken at the request of Headquarters, United States Air Force (SC), the Assistant Secretary of the Air Force for Acquisition (SAF/AQS), and Air Force Space Command (XP and SC). It is part of the Employing Commercial Communications task within Project AIR FORCE's Aerospace Force Development Program.

This research should be of interest to those concerned with obtaining satellite communications in the Air Force, the other military services, and the defense agencies.

## PROJECT AIR FORCE

Project AIR FORCE, a division of RAND, is the Air Force federally funded research and development center (FFRDC) for studies and analyses. It provides the Air Force with independent analyses of policy alternatives affecting the development, employment, combat readiness, and support of current and future aerospace forces.

Research is performed in four programs: Aerospace Force Development; Manpower, Personnel, and Training; Resource Management; and Strategy and Doctrine.

# CONTENTS

# TABLES

The Department of Defense (DoD) is considering major investments in systems that exploit information to support warfighting. Communications among users around the globe is key to transmitting and using this information, and currently programmed DoD systems will not satisfy the total projected communications demand with dedicated military assets. In this study, we seek to answer four sets of questions:

- How much of the projected demand can be met with programmed and planned military assets?

- Can commercial technologies, systems, or services meet the remaining needs? How do commercial communication assets compare with military assets in their ability to meet criteria important to DoD? What steps might be taken to mitigate shortfalls?

- What is the expected cost of providing the projected communications demand?

- What investment strategies should DoD employ to minimize the expected cost?

We examined a specific category of communications—high-bandwidth, minimally protected satellite communications. Although this is a subset of total military communications demand, it represents roughly half of the projected military capacity needs. If commercial systems can satisfy this need, military systems can be used for communications requiring more specialized characteristics or greater levels of control over their operation.

## DEMAND AND SUPPLY

DoD projects routine, day-to-day demand for long-haul, wideband military communications to grow from 1 gigabit per second (Gbps) today to roughly 9 Gbps in 2008 (see Figure S.1). Projected surge demand ranges from less than 1 Gbps now to approximately 4 Gbps in 2008. Total demand is thus projected to grow from 1 Gbps now to almost 13 Gbps in 2008. The current capacity of military satellites capable of providing wideband communications is on the order of 1 Gbps. That level will begin growing in 2004 as the Gapfiller satellite system comes into operation but will stay below 4 Gbps under

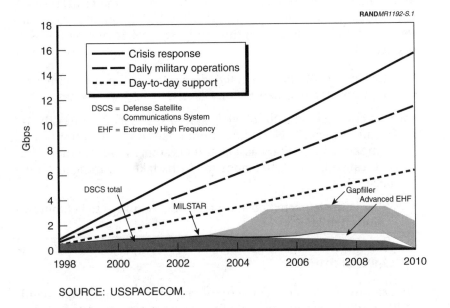

Figure S.1—Comparison of Projected Military Demand and
Programmed Supply

current plans. Supply of communications capacity by DoD assets thus already falls short of potential military demand, and that shortfall may grow to more than 8 Gbps by 2008.[1]

Commercial systems supply between 200 and 250 Gbps of long-haul, wideband capacity to a variety of commercial and governmental users. Commercial supply, however, is unevenly distributed geographically, and it is not certain that sufficient capacity will always be immediately available to meet military demand unless arranged in advance. DoD should seek commercial lease contracts that allow switching capacity allotted to DoD on any given satellite from one transmitting beam to another (or moving the beam itself). This will permit flexibility in coverage. There is no wideband commercial coverage over the North or South Poles, and it is uncertain that any will be provided in the foreseeable future.[2]

## ABILITY OF COMMERCIAL SYSTEMS TO MEET OTHER MILITARY CRITERIA

Although commercial systems can easily provide the capacity DoD needs and can match current DoD assets in coverage, some in DoD have questioned the ability of commercial systems to meet other criteria. We have assessed that ability for several criteria and have reached the following conclusions:

- **Flexibility.** Lease of commercial transponders or services should enhance flexibility in terms of access to a variety of frequencies and locations around the world. We see an advantage in employing commercial communications in this regard.

- **Interoperability.** Because most satellites simply transmit information without processing it, interoperability applies principally to terminal equipment. DoD can in theory maintain

---

[1]The fact that DoD-owned capacity falls short of current military demand is evidenced by the large amount of commercial capacity already leased by DoD. According to United States Space Command (USSPACECOM), that amount today is over 1 GHz of commercial bandwidth, which we estimate is capable of carrying 900 Mbps of digital traffic.

[2]The Iridium system does provide narrowband capacity over the poles. Some of the new low earth orbit (LEO) systems proposed might provide additional capacity.

interoperability among its own terminal equipment, but it has less control over commercial systems. In the case of future commercial processing satellites, nobody knows which systems will achieve mass use. DoD might mitigate this problem by encouraging the development and adoption of commercial protocol standards, and by designing into its equipment the ability to upgrade cheaply as technology advances.

- **Access and Control.** DoD should not have any problems gaining access to or controlling the operations of a satellite for which it has exclusive rights of use. That applies to satellites DoD leases from commercial owners as well as to those it owns. In the case of fractional-satellite leases, however, DoD naturally cedes unlimited rights of access and control. It may not automatically have the right to reorient its coverage by switching its allocated capacity to a different radiating beam. And it may be accorded no priority over another customer by an end-to-end service provider. Clearly, *some* commercial satellite options have disadvantages with respect to this criterion—in certain cases very large ones relative to DoD-owned assets.

- **Quality of Service.** Commercial systems should be able to meet any reasonable standard of quality and transmission reliability. The global communications system is also robust enough so that breaks in service should be quickly restored by rerouting messages through alternate channels. With both military and commercial systems DoD will need to ensure that it has bought the level of reliability needed for each application.

- **Protection.** Current and planned commercial communications satellites are designed to achieve optimal utilization of bandwidth at minimal cost within a benign environment. DoD has stated that current commercial satellites and DoD-owned "commercial-like" satellites offer only minimal protection against intentional interference, detection, or interception. Half of the capacity demand projected requires no protection; for the remainder, DoD has stated that specially designed satellites will be needed to provide the protected communications desired.

If DoD is to meet its satellite capacity needs for the near future, it will have to use some amount of commercial communications. The compromises it may have to make in the characteristics listed above may not be very important for some purposes; for others, they may be preferable to forgoing the capacity. In addition, it may be possible to take steps that will substantially reduce the disadvantages of using commercial systems.

## CHOOSING INVESTMENT STRATEGIES

DoD has a range of choices for procuring commercial satellite capacity. It can buy satellites or lease them. If it buys, it can purchase more capacity than needed for day-to-day demand to accommodate contingencies, or it can wait and purchase only enough capacity to satisfy demand as it emerges. If it leases, it can engage in short-term contracts ranging from a week to a year or it can commit to longer leases on the order of one to ten years. The various approaches can be mixed.

We consider first a choice between the following two approaches:

- Buying a satellite, or committing to long-term leases only when sufficient day-to-day demand has emerged, and making up any capacity shortfalls arising from contingencies with short-term (less than one year) leases.

- Buying more satellites, or committing to more long-term lease capacity, than required so that the need for subsequent short-term leases is reduced.

Capacity procured ahead of demand (by buying satellites or leasing capacity) is less expensive than leasing over short terms. In fact, because short-term leases are relatively expensive compared with long-term procurement commitments for the same capacity, acquiring extra (or "slack") capacity can (up to a point) actually save money.

One-year leases are less expensive (per year) than one-week leases, and ten-year leases are less costly still. We show that buying satellites can be less expensive over the long run than even ten-year leases. However, the premium paid to lease capacity, rather than buying satellites, decreases when there is a gap between the order and receipt of a satellite because DoD must make payments for the satellite as it is built—before the expected annual savings over leasing are realized. The effect of an order-receipt lag on the ratio of leasing costs to buying costs is given in Table S.1.

A one-year lease would cost 52 percent more than the annual cost of a satellite purchased with no lag between order and receipt; a ten-year lease would cost 26 percent more than the annual cost of purchasing that satellite. With a three-year lag between the order and receipt of a purchased satellite, the premium for one-year leases drops to 16 percent, and the ten-year lease becomes cheaper than purchasing. With a five-year lag, the one-year lease is 4 percent more expensive than purchasing, and the ten-year lease price drops to 86 percent of the price of purchasing.

## CONCLUSIONS

Our analysis leads us to five general conclusions regarding the employment of commercial satellite capacity by DoD:

- DoD projects a large gap between its demand for communications and the capacity expected from present, planned, and programmed systems. Its options are to limit the amount of communications available to users, buy more DoD-unique systems, or employ commercial systems. Commercial leases provide a valuable way to increase capacity even when DoD buys unique systems.

- Cost is not the only criterion—sometimes DoD needs to pay more for military-unique operational capabilities. Where access, control, and protection are high priorities, DoD will have to rely on its own assets. Where these characteristics are not truly essential, DoD should be able to use commercial capacity while

**Table S.1**

**Ratio of Lease Cost to Purchase Cost for 1 Gbps of Capacity**

| Contract Terms | Lease/Purchase Cost Ratio, No Order-Receipt Lag | Lease/Purchase Cost Ratio, 3-Year Order-Receipt Lag | Lease/Purchase Cost Ratio, 5-Year Order-Receipt Lag |
| --- | --- | --- | --- |
| One-year lease | 1.52 | 1.16 | 1.04 |
| Ten-year lease | 1.26 | 0.96 | 0.86 |

taking steps to improve access to, control over, and protection of these systems to the levels needed. Lease contracts should include the rights to switch transponders between beams as needed.

- DoD must develop operational concepts that maximize its flexibility in employing commercial and DoD systems in order to meet both day-to-day and contingency demand in a way that does not make it unduly vulnerable to enemy disruption, technical failures, or market forces.

- It may be more economical to make long-term commitments and "waste" some capacity than to underestimate need and make up the shortfall with short-term service contracts. It will depend on the expected demand and the long-/short-term price ratio.

- A three-year lag between the order and receipt of a DoD-unique satellite reduces the premium for a ten-year lease to zero, and a five-year lag results in near parity between purchasing a satellite and leasing the same capacity with a series of one-year leases. Therefore, expected savings should not motivate buying DoD-unique satellites. DoD should make the choice based on the operational characteristics needed.

# ACKNOWLEDGMENTS

The authors are indebted to our sponsors for supporting this study and providing comments, suggestions, and additional information as we progressed. The sponsors are Lt Gen William Donahue (AF/SC), Lt Gen Woodward (formerly AFSPC/SC, now Joint Staff J-6), Brig Gen James Beale (ret.) and Brig Gen Clay (SAF/AQS), and Brig Gen John Boone (ret.) and Brig Gen Donald Petit (AFSPC/XP). We are also indebted to their staffs within the Air Force and United States Space Command for their active participation throughout our effort. Of special note are Col Hal Hagemeier, Col John Payne, Lt Col Milton Johnson, Lt Col Ken White (ret.), Lt Col Dave Spataro, Lt Col Denise Knox, Lt Col Mike Sinisi, Maj Cynthia DeCarlo, and Maj Lance Spencer.

Our colleagues at RAND aided significantly in this effort. Brent Bradley gave us the initial charter to begin work in this area. Natalie Crawford helped focus our efforts upon the fundamental problems the Air Force and the Department of Defense are facing, and provided intellectual leadership. Glenn Buchan and Charlie Kelley participated in numerous briefings at RAND and with the client. Willis Ware and Dick Neu provided background information and strengthened the rigor of our analyses. Dick Neu also served as a reviewer of the final document and helped in focusing the final product. Joel Kvitky contributed to our technical analyses and arguments, and also served as a reviewer. Any remaining errors or omissions are, of course, the responsibility of the authors.

| | |
|---|---|
| Advanced EHF | Advanced Extremely High Frequency |
| AFB | Air Force Base |
| AOC | Air Operations Center |
| AOR | Area of Regard |
| ATM | Asynchronous Transfer Mode |
| CENTCOM | Central Command |
| COMSAT | Communications Satellite |
| CONUS | Contiguous United States |
| CRD | Capstone Requirements Document |
| CSCI | Commercial Satellite Communications Initiative |
| CVBG | Carrier Battle Group |
| DII | Defense Information Infrastructure |
| DISA | Defense Information Systems Agency |
| DISN | Defense Information Systems Network |
| DoD | Department of Defense |
| DoDOSA | Department of Defense Office of Space Architect |
| DSCS | Defense Satellite Communications System |
| DWTS | Digital Wideband Transmission System |
| EELV | Evolved Expendable Launch Vehicle |
| EHF | Extremely High Frequency |
| EIRP | Effective Isotropic Radiated Power |

| | |
|---|---|
| EMP | Electromagnetic Pulse |
| ERDB | Emergency Requirements Data Base |
| FCC | Federal Communications Commission |
| GAO | General Accounting Office |
| Gbps | Gigabits Per Second |
| GBS | Global Broadcast System |
| GEO | Geo-Synchronous Earth Orbit |
| GHz | Gigahertz |
| HQ | Headquarters |
| IEEE | Institute for Electrical and Electronics Engineers |
| IP | Internet Protocol |
| ITU | International Telecommunication Union |
| JNMS | Joint Network Management System |
| JROC | Joint Requirements Oversight Council |
| JTF | Joint Task Force |
| JTIDS | Joint Tactical Information Distribution System |
| Kbps | Kilobytes per Second |
| KHz | Kilohertz |
| Km | Kilometer |
| LEO | Low Earth Orbit |
| LMST | Lightweight Multiband Satellite Terminal |
| LOS | Line of Sight |
| LPD | Low Probability of Detection |
| LPI | Low Probability of Intercept |
| Mbps | Megabits Per Second |
| MDR | Medium Data Rate |
| MEO | Medium Earth Orbit |
| MHz | Megahertz |
| MILSATCOM | Military Satellite Communications |
| MILSPEC | Military Specification |
| MILSTAR | Military Strategic and Tactical Relay |

| | |
|---|---|
| MTW | Major Theater War |
| NASA | National Aeronautics and Space Administration |
| NATO | North Atlantic Treaty Organization |
| NCA | National Command Authority |
| O&S | Operations and Support |
| OECD | Organization for Economic Cooperation and Development |
| RDT&E | Research Development Test and Evaluation |
| SAF | Secretary of the Air Force |
| SATCOM | Satellite Communications |
| SHF | Super-High Frequency |
| SINCGARS | Single Channel Ground and Airborne Radio System |
| SINR | Signal-to-Noise Ratio |
| SIOP | Single Integrated Operational Plan |
| SLEP | Service Life Extension Program |
| SSC | Small-Scale Contingency |
| STAR | System Threat Assessment Report |
| STAR-T | Triband Advanced-Range Extension Terminal |
| STEP | Standardized Tactical Entry Point |
| SWA | Southwest Asia |
| SWARF | Senior Warfighter Forum |
| TWCF | Transportation Working Capital Fund |
| UAV | Unmanned Air Vehicle |
| UFO | Ultra-High Frequency Follow-On |
| UHF | Ultra-High Frequency |
| USAF | United States Air Force |
| USSPACECOM | United States Space Command |

# Chapter One

# INTRODUCTION

The Department of Defense (DoD) is considering major investments in systems that exploit information to support warfighting, and communications between users around the globe will be key to transmitting and using this information. In the near term, there are not enough military systems to satisfy projected communications demand and commercial systems will have to be used. In the future, budgetary pressures will make it difficult for the services to satisfy the projected communications demand with dedicated military assets.

In this report, we seek to answer several questions:

- How much of the projected demand can be met with programmed and planned military assets?

- Can commercial technologies, systems, or services meet the remaining needs? How do commercial communication assets compare with military assets in their ability to meet criteria important to DoD? What steps might be taken to mitigate shortfalls?

- What is the expected cost of providing the projected communications demand?

- What investment strategies should DoD employ to minimize the expected cost?

The many categories of military communications include everything from battlefield communications between mobile users to communications between fixed sites in rear areas. Some of these communications must be survivable in a nuclear war, and others need high levels of protection from detection, interception, or jamming. Some require very high data rates, whereas others need only low data rates. Some communications can be by wire or fiber optic cable, whereas others must use wireless means.

We have examined a specific category of communications—high-bandwidth, minimally protected satellite communications. This category of military demand represents roughly half of the projected military satellite capacity needs. To the extent that use of commercial systems can satisfy this need, military systems can be used for more specialized communications needing a greater level of control over their operation.

## DoD DEMAND FOR COMMUNICATIONS AND THE ROLE OF COMMERCIAL SATELLITES

The ability to communicate is fundamental to military activities—providing information to field commanders, commanding and controlling forces, and sending targeting information to combat units. Military strategy doctrine, theory, and rhetoric are increasingly occupied with information and its potential for improving combat performance.[1] Transmitting this information will require improvement in the technology and capacity of wideband communications.

The DoD and the analytic community expect military demand for communications to grow over the next decade and beyond. Some in the military have reported a seven-fold increase in demand over the last decade; others project a similar growth in the next.[2] Sufficient capacity for transmitting information must be obtained to support emerging military doctrine, including the uncertainties posed by the

---

[1] *A National Security Strategy for a New Century* (Clinton, 1998) states that "improved intelligence collection and assessment coupled with modern information processing, navigation and command and control capabilities are at the heart of the transformation of our warfighting capabilities." *The National Military Strategy* (Joint Chiefs of Staff) adds that "Information superiority allows our commanders to employ widely dispersed joint forces in decisive operations, engage and reengage with the appropriate force, protect the force throughout the battlespace, and conduct tailored logistical support." *Joint Vision 2010* (Shalikashvili, 1997a) states that improvements in "information and systems integration technology" is one of the technology trends expected to provide "an order of magnitude improvement in lethality."

[2] See Raduege (1998).

unknowable timing of future contingencies. This raises some important questions: How can DoD acquire more communications? Can DoD afford to buy enough DoD-unique systems? Can DoD obtain some portion of the needed capacity from commercial systems?

In theory, the military could simply build enough DoD-unique communications satellites to satisfy the projected demand. However, there are many competing needs within DoD for the DoD budget. The decisions of how much budget should be spent on specialized DoD communications assets and how much capacity should be obtained commercially go well beyond the Air Force. Present guidance is to limit spending on satellite communications to the amount budgeted in the 1998 President's Budget,[3] which will make it difficult to satisfy projected communication needs with military-owned and -unique systems.

One possible solution may be to employ commercial communications to augment military satellite capacity and coverage (see Chapter Three for the current global communications market and satellite acquisition alternatives). As we show in Chapter Four, present capacity and proposed expansions dwarf projected increases in military capacity needs. Thus, commercial systems and services may represent the best opportunity to achieve affordable communications capacity .

## MEETING MILITARY CRITERIA WITH COMMERCIAL SYSTEMS

Each service and combatant commander must decide what combination of capacity, coverage, quality of service, flexibility, interoperability, access and control, and protection is needed, and must weigh each of these characteristics against the cost required to achieve it. DoD today uses commercial communications for many applications. Some of these applications may require specialized systems, and if commercial customers see advantages in these systems as well, then they may share in the investment needed for their development.

---

[3]See HQ USSPACECOM (April 24, 1998).

When control and protection found in commercial systems can be accepted, those systems may be viable alternatives. Military operational concepts should include available commercial systems and ensure that their use does not make military operations unduly vulnerable to enemy actions such as jamming. In Chapters Five and Six, we evaluate the ability of commercial systems to meet a range of military needs and propose some notional operational concepts to mitigate the effects of enemy actions. In Chapter Seven, we estimate the steady-state costs of obtaining capacity through commercial leases and from DoD-owned satellites.

## DoD DEMAND VARIANCE AND ALTERNATIVE INVESTMENT STRATEGIES

The task of planning a communications investment strategy would be simple if demand were known with certainty. Indeed, many proposed strategies attempt to ignore the uncertainty in future demand. Unfortunately, the variance in DoD demand arising from contingencies is large, as we will show in Chapter Eight. Military investment strategies must deal effectively with this variance and with uncertain long-term growth. In Chapter Nine, we evaluate the cost of several alternative strategies to meet DoD capacity needs in the presence of variance in demand and uncertain demand growth.

# Chapter Two

# DoD COMMUNICATIONS DEMAND PROJECTIONS

In this chapter, we discuss estimates of future military communications demand. After providing some context, we use an established DoD projection as a starting point for analysis. The projections given in this chapter are steady and do not take into account uncertain events such as contingencies. In Chapter Eight, we will show how uncertainties arising from the timing of contingencies can result in substantial variations from any smooth projection line.

## CONTEXT

The Air Force and DoD initiated satellite communications (SATCOM) developments in the 1960s. Since that time, SATCOM has become an integral part of military operations—from transmitting a common operational picture to allowing rear-area units to perform otherwise-impossible logistical and intelligence functions. We have obtained two types of information regarding expected demand for satellite communications:

- Statements of security and military strategy, and more detailed descriptions of doctrine.[1] The treatment of communications in these documents is far more qualitative than quantitative.

- Quantitative projections of communications demand contained in such documents as the Advanced Military Satellite Communications Capstone Requirements Document (hereafter

---

[1]See Clinton (1998) and Shalikashvili (1997a, b), among others.

referred to as the Capstone document or CRD),[2] which will be reviewed in the following sections.

From these documents, several general insights can be gained:

- The military is fundamentally rethinking its operations and developing doctrine intended to increase mission performance, decrease casualties, reduce the time and forces needed to achieve objectives, and decrease the costs of operations.[3]

- To enable this doctrine, military commanders need timely access to a vast amount of intelligence, surveillance, and reconnaissance information. This information will need to be communicated rapidly throughout the theater and across the world.

- Senior political and military leaderships do not want communications to constrain operations. That is, there is an implied expectation that the military will have access to whatever type and amount of communications it requires to support operations.

## DoD PROJECTIONS OF FUTURE MILITARY DEMAND

The Capstone document and related documents are a point of departure for our quantitative analysis. The CRD was "developed to provide the 'capstone' requirements framework and operational concept on which to develop an affordable architecture and acquisition course of action." The Capstone projections used in this chapter focus on wideband satellite communications[4] because wideband

---

[2]The Capstone demand projections are built from the Emerging Requirements Data Base maintained by the Joint Staff's Directorate of Command, Control, Communications, and Computers. See HQ USSPACECOM (1998).

[3]"The information we receive from and through space will be a key enabler of the four operational pillars of *Joint Vision 2010*: Dominant Maneuver, Precision Engagement, Focused Logistics, and Full-Dimension Protection." For complete text, see Estes (1998).

[4]Wideband refers to services having channels greater than 64 kilobits per second (Kbps). Narrowband refers to services having channels with less than 64 Kbps. Narrowband channels typically provide voice services and specialized data services not needing high data rates. Because the total aggregate capacity of DoD and commercial narrowband systems is small, they are not discussed in this analysis.

service represents the bulk of present and future DoD demand for satellite communications. Wideband service also makes up a large segment of the commercial market, making it attractive for identifying potential opportunities.

The Capstone document provides two sets of projections. The first comes from the Emerging Requirements Data Base (or ERDB), and the second was generated by United States Space Command (USSPACECOM). The ERDB demand projection is shown in Figure 2.1 as gigabits per second (Gbps) capacity as a function of year. The ERDB has been validated by the Joint Requirements Oversight Council (ROC) for use in planning the amounts of military satellite communications (MILSATCOM) services U.S. forces are projected to need in the 2010 timeframe. The ERDB demand projection is disaggregated into three major components: (1) day-to-day support activities, (2) daily military operations, and (3) crisis response.

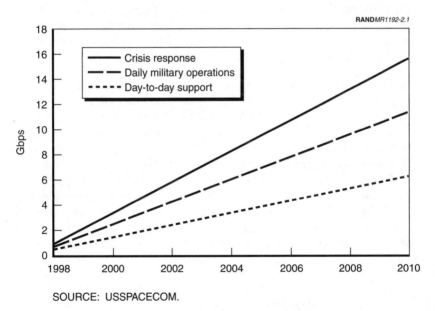

SOURCE: USSPACECOM.

**Figure 2.1—ERDB Projections of MILSATCOM Capacity Demands**

USSPACECOM explained the day-to-day support category as representing the minimum capacity needed to ensure vital peacetime communications. These include indications and warning of attack against the United States or interests abroad, command and control of U.S. nuclear forces, communications between the national command authority and combatant commands, and support to intelligence agencies and the diplomatic telephone service. In addition, day-to-day support includes some of the capacity of the Defense Information Systems Network (DISN) "backbone"—the communications network tying together all DoD users.

Daily military operations include routine patrols, tactical intelligence, and training and exercises. It also includes a category that the Capstone document calls the "current crisis"—those contingency operations to which forces might be already committed at any given time. Crisis response includes the additional capacity needed to respond to multiple (up to four simultaneous) small-scale contingencies or two major theater wars. These communications support in-transit and deployed forces, and both in-theater communications and "reachback" communications to out-of-theater locations.[5]

USSPACECOM developed its own estimates of day-to-day and contingency communications needs and included them in the Capstone document. The USSPACECOM estimate of total day-to-day demand in the 2005 through 2010 timeframe is approximately 9300 megabits per second (Mbps—millions of bits per second);[6] the total demand for a small-scale contingency is approximately 1700 Mbps;[7] and the total demand for a major theater war is about 3900 Mbps.[8] The Capstone day-to-day demand of 9300 Mbps is roughly equivalent to the ERDB total of day-to-day support activities and daily military operations in 2008. The Capstone total of 3900 Mbps for a major theater war is very close to the ERDB quantity of 4000 Mbps for crisis response in 2008.

---

[5]Although many of the individual communications included in these categories are narrowband we focused our analysis on the subset of these communications that use wideband services, which use the dominant portion of projected capacity demand.

[6]See HQ USSPACECOM (1998), pp. 4–17.

[7]*Ibid.*, p. H-6.

[8]*Ibid.*, p. H-7.

The current capacity of military satellites capable of providing wide-band communications is on the order of 900 Mbps. This includes the Defense Satellite Communications System (DSCS), the Global Broadcast System (GBS), and Military Strategic and Tactical Relay (MILSTAR). DoD-owned capacity will begin growing in 2004 as the Gapfiller satellite system comes into operation. However, total DoD-owned capacity will stay below 4 gigabits per second (Gbps—billions of bits per second) unless additional systems are acquired. DoD communications capacity thus already falls short of potential military demand, and that shortfall may grow to more than 8 Gbps by 2008 (see Figure 2.2).[9]

There are two further points regarding the use of demand projections in our analysis:

- It is impossible to forecast demand precisely, and we neither accredit nor imply validation for the projections we show here.

- These projections do not describe the variance in demand that should be anticipated.

Demand projections depend on the type of operations expected, operational tempo, size and type of deployed forces, and fielded technologies. In addition, communications needs at each force level grow as technologies evolve and become increasingly integrated into tactical operations. The CRD projections consider these factors, and attempt to bound the range of demand that should be anticipated during peacetime and crisis. Further, although the Capstone projections do provide information on uses of communications that will drive future demand, these projections cannot yield precise numbers for yearly demand or demand growth.[10] There has been substantial

---

[9]DSCS is an X-band military satellite system with some protected capacity. The capacity shown in Figure 2.2 includes the higher-capacity Service Life Extension Program (SLEP) satellites. MILSTAR satellites use extremely high frequency (EHF) signals and are highly protected and survivable. Capacity shown in the figure includes the MILSTAR II medium data rate (MDR) payload. The Advanced EHF satellites are planned to be high-capacity, highly survivable and protected replacements for MILSTAR. See Appendix A for a description of frequency bands, and Appendix B for a description of current military communications satellites.

[10]Coordination among communicators helps to optimize the use of existing capacity but can lead to underrepresentation of future needs. Users typically only report needs they expect to be supported by existing SATCOM resources. Unidentified—but real—

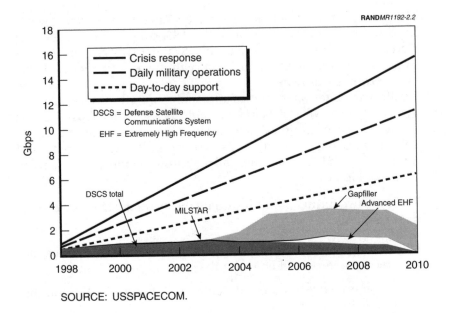

SOURCE: USSPACECOM.

**Figure 2.2—Comparison of Projected Military Demand and
Programmed Supply**

demand variance within a year depending upon the timing, size, and duration of contingencies; this variance must be dealt with to minimize expected costs. The Capstone projections of future military demand will be used for comparative purposes in Chapters Three and Four. The variance in communications demand arising from contingencies will be discussed in Chapter Eight.

---

needs may emerge and compete with other low-priority needs as advances in SATCOM capacity and technology change user expectations and capabilities. The CRD projection does not explicitly account for such potential unidentified needs.

# Chapter Three

# MEETING MILITARY DEMAND WITH COMMERCIAL SATELLITES: CONTACT AND EVALUATION

Commercial communications have undergone extraordinary growth, and the advent of the Internet, new consumer services such as direct-to-home satellite television, and international agreements to open telecommunications markets offer many opportunities to develop profitable new services.[1] In Chapters Four, Five, and Six we will examine the various dimensions of communications supply from commercial satellites. First, we review international cable and satellite systems and their applications in the communications market. Second, we set out the satellites and network systems we intend to evaluate and the criteria against which we evaluate them.

## CONTEXT: INTERNATIONAL COMMERCIAL COMMUNICATIONS SYSTEMS

### Terrestrial Systems and Networks

Earthbound transoceanic communications are carried by undersea cables that connect most of the world's principal cities. Revolutionary advances in commercial submarine cable systems now permit enormous capacity between major telecommunications hubs. In 1988, the first-generation fiber-optic cable—TAT 8—was laid, with the capacity for 22,000 simultaneous voice calls.[2] By 1993,

---

[1]These agreements include the 1997 World Trade Organization agreement to open national telecommunications markets to foreign competition.

[2]Assuming each call is allocated 12.8 Kbps.

second-generation systems, with far more capacity, were being laid—TAT 12/13 across the Atlantic and TPC 5 across the Pacific. These systems can transmit 5 Gbps on two fibers, with an additional 5 Gbps of capacity in reserve in case of failures in the main cables. The latest transatlantic and trans-Pacific cables have been designed to employ multiple wavelengths or "colors" simultaneously on each fiber, with each "color" providing 2.5 Gbps capacity.[3] These cables have been laid in optical network rings that can route calls around breaks in the cables without interrupting communications. Over the next several years, new cables with capacities ranging from 40 to 100 Gbps or more are planned across the Atlantic and Pacific oceans. Figure 3.1 illustrates some of the systems that have been deployed, and notes their capacities on a logarithmic scale (with each major division signifying an order of magnitude increase). The capacity of these cables has far surpassed that of large-capacity satellites in service today.

Sufficient capacity is typically available on these cables for the military to use without owning them. However, cables do not provide service to remote areas or for mobile operations. Microwave relays may be useful in some cases, but require time to set up and must be protected. Communications relays on unmanned air vehicles (UAVs), useful for some applications, may be vulnerable to enemy air defenses and require time to set up the infrastructure for flight operations. Hence, the military has a continuing need for satellite communications.

## Satellite Systems and Applications

The global satellite communications market has grown remarkably since the first commercial satellite attained full operational capability in 1965. That satellite, Early Bird,[4] was used to carry voice and television traffic between the United States and Europe. Early Bird capitalized on many design features of the Syncom satellites first

---

[3]Not all of this capacity need be "lit" upon installation. Often, one or two "colors" are "lit" at commencement of service, with equipment added to the terminal ends to light additional colors as needed. This has the benefit of making capacity upgradeable without changing the submerged cable.

[4]Also known as INTELSAT I (see Martin, 1996).

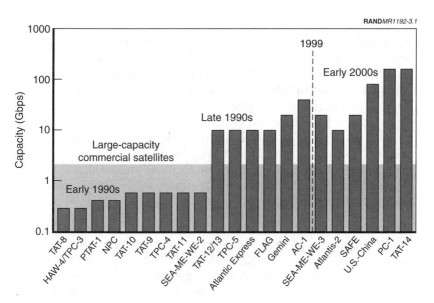

SOURCE: Federal Communications Commission.

**Figure 3.1—Major Submarine Cables in Use and Under Development**

launched for the DoD in 1963. Domestic satellite communications began in 1974 with the launch of Westar I. Satellite weight, power, transponder[5] number and capacity, and the number of fixed and movable beams have increased steadily over the years.

Commercial communications satellites today are in low earth orbits (LEOs), medium earth orbits (MEOs), and geosynchronous earth orbits (GEOs). Iridium and Globalstar are examples of current LEO systems, and the proposed Teledesic system is another.[6] Most commercial communications satellites are in geosynchronous earth orbit—at 35,700 km altitude.

---

[5]Transponders, also referred to as repeaters, receive, amplify, and retransmit signals between users at separate ground stations. They are the "working" or moneymaking part of the satellite.

[6]Their orbital altitudes are 770 km, 1400 km, and (planned to be) 1300 km, respectively.

In geosynchronous earth orbit, the satellites appear to be stationary when viewed from the earth.[7] In general, this is seen as an advantage, because a ground antenna can be pointed at the satellite once and remain in contact. The disadvantage is that it takes between one-quarter and one-half of a second to travel to the satellite and back. This delay time—or latency—is noticeable in voice conversations. There has also been concern that latency could prove troubling for Internet protocols that must receive confirmation of packet receipt.[8] Satellites at LEO and MEO have less latency, but require that a constellation of satellites be built to ensure that any given point on the ground always has at least one in view. This adds to the cost of the space segment.

We will not consider LEO satellite constellations further in this report for two reasons. First, only three such satellite constellations, Iridium, Globalstar, and ORBCOMM, exist; the remainder are proposed at this point. Second, these are narrowband (low-data-rate) systems intended primarily for paging or voice. The focus of our research has been on wideband systems capable of sending voice, video, and data. If a wideband system such as Teledesic is deployed, its use should be considered by the military. We will note the effect such a system might have on commercial capacity in the next chapter.

Prior to launch, satellites go through a lengthy coordination process with the International Telecommunications Union (ITU) and the nations that are in view of the satellite. The ITU serves as an arbiter and administrator for those nations that have agreed to coordinate their use of the communications spectrum with others.[9] This includes virtually all nations. However, the ITU has reported some instances of noncompliance with ITU guidelines. The major penalty for not coordinating use of the spectrum is interference with other commu-

---

[7]In fact, some stationkeeping is needed to maintain position. When fuel runs out, the satellite begins to develop a "figure 8" pattern or a "wobble," necessitating the use of more complex and expensive ground antennas to track the satellite.

[8]The concern is that packets would be assumed lost because of the delay in confirmation. Packets would be resent, again be assumed lost, and be resent in an endless loop, eventually bogging down the system. The emergence of "Internet protocol (IP) over satellite" suggests the problem is manageable.

[9]Nations, not individuals or companies, file with the ITU for spectrum assignments.

nications users. Most nations, therefore, see it as in their best interest to coordinate their spectrum usage[10] to ensure that a satellite in a specific orbital position (or "slot") will not interfere with previously established users of a given frequency band. In addition, coordination establishes a satellite as the acknowledged user of a specific band in a slot.[11]

The ITU will allocate spectrum for specific orbital slots to those nations filing on a first-come/first-served basis for the useful life of the satellites, without respect to whether these systems will be used by commercial or government entities. The filing nations then coordinate their intended use of these frequencies with the nations that will be in view of the satellite. The nations in view of the satellite can raise objections if use of these frequencies by satellites will interfere with other uses within their borders. If the coordination process is successfully completed, then the respective nations agree to allow satellite operations to commence. However, this does not grant the nation operating the satellite any implied landing rights—approval to receive or transmit signals must be obtained separately from each host-nation government.

Commercial satellites carry voice, video, and data between fixed and some mobile users. Satellites have transponders and a "bus" that performs "housekeeping functions" such as providing power, control, and stationkeeping. Transponders (repeaters) receive, amplify, and retransmit signals. Today, commercial satellites use C-band transponders with global, hemispheric, and zonal beams, and Ku-band transponders with spot beams to provide connections between users (Figure 3.2).

Global and hemispheric beams are used for applications (such as telephony) requiring access to broad regions. Zonal and spot beams

---

[10]International Telecommunications Union (1997).

[11]Some orbital slots are claimed but vacant. This can happen when a nation has filed an intention to commence services with the ITU, but has not yet bought a satellite. These are sometimes referred to as "paper satellites." Though contrary to ITU agreements, this may be a way to claim a piece of spectral real estate for future use or sale to others.

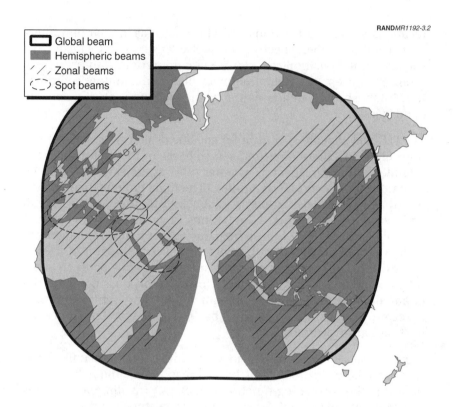

RAND*MR1192-3.2*

Figure 3.2—Exemplar Satellite Beam Patterns

provide focused coverage for applications such as video and data.[12] A satellite might have each type of beam, with transponders using the same frequencies in several beams simultaneously. Interference is avoided by physically separating and polarizing the beams.

To discriminate between satellites using the same frequencies, the satellites are spaced 2 degrees apart; this implies that only 180 satellites using each band could be in geosynchronous orbit at any one

---

[12]Reducing the size of the spot concentrates transponder power. For example "direct-to-home" television uses high-powered transponders in spot beams in small home dishes.

time.[13]  Existing and planned satellites, if all were in operation, would more than fill the available slots.[14]  However, many will be used as spares or retired once new satellites are launched.  Also, some companies have begun to locate several satellites (sometimes five or more) in the same orbital slot.[15]

In addition to C- and Ku-bands, S-, L-, and Ka-bands are of interest to commercial users.  The S- and L-bands are primarily used for mobile telephony (e.g., to ships at sea) and messaging. Although Ka-band has been experimented with for some time, technical challenges have delayed commercial use.[16]  Ka-band is attractive because it has been assigned a large bandwidth by the FCC—making "broadband" or high-data-rate services possible.  Also, the first groups filing for allocations were able to propose systems without having to coordinate with established users.[17]  Some military satellites use X-band frequencies with low susceptibility to atmospheric attenuation and limited commercial use[18]—meaning less competition for orbital slots.

## SATELLITE ALTERNATIVES AND EVALUATION CRITERIA

The military has several options to match supply and demand for military communications:

---

[13]Advances in technology, higher frequencies, and precise coordination of beams may allow closer spacing in the future.

[14]Currently, there are 141 satellites in geostationary orbit broadcasting at C-band and 154 at Ku-band.  An additional 59 C-band and 93 Ku-band are planned for launch over the next four years.

[15]Direct-to-home video broadcasters often place several satellites in a "hot bird" slot to accommodate a large number of homes or businesses with receive-only dishes.

[16]Ka-band is used on the National Aeronautics and Space Administration (NASA) Advanced Communications Technology Satellite (ACTS) and on Japanese test satellites.  However, Ka-band signals suffer more than X- or Ku-band from degradation caused by rain.

[17]On May 9, 1997, the International Bureau of the Federal Communications Commission (FCC) awarded licenses for 13 proposed commercial Ka-band satellite systems.

[18]Frequencies between 240 and 340 MHz (ultra-high frequency, UHF), 7250 and 7750 MHz and 7900 and 8400 MHz (X-band), and 19.2 and 20.2 GHz and 29 and 30 GHz (Ka-band) are recognized for exclusive military use by the FCC within the United States.

- DoD could attempt to limit the amount of communications that U.S. forces use to match the amount that present, programmed, and planned military systems can provide;

- DoD could allocate more money for purchasing military-unique communications satellites;

- DoD could employ additional commercial communications to augment military systems.

We do not consider the first option further. Although it is unclear what future contingency needs will be, restricting the communications available to U.S. forces may make it difficult to have the real-time command and control envisioned in joint doctrine. The second option requires that more money be allocated to buy communications satellites—money that would need to come from an increase in DoD funding, a shift from other acquisition programs, or a shift from other accounts such as operations and support (O&S).

The third option requires more money, too. In addition, it requires DoD to develop concepts to employ systems that the military does not completely control. Currently, DoD appears to be satisfying about half of its satellite communications needs with commercial systems. Even if more military satellites are purchased, commercial systems would provide a valuable option for additional communications to meet demand growth or unexpected contingency surges.

In concept, the choices facing DoD are not only between military and commercial systems but also between owning and leasing (see Table 3.1). Ownership includes all rights and obligations pertaining to the hardware itself during and after its projected useful life and all rights of use for the same period. In addition, ownership requires an orbital slot with the appropriate spectrum allocation. Leases can include the right to use a specified satellite, transponder, or bandwidth for a specified amount of time. The hardware or service leased typically occupies or originates from a slot on which the owner has an established claim.

**Table 3.1**

**DoD Satellite Communications Acquisition Options**

| Option | Military-Unique System[a] | Commercial System[a] |
|---|---|---|
| DoD ownership | DSCS MILSTAR UFO Gapfiller | Policy difficulties |
| DoD lease | LEASAT | INTELSAT PanAmSat Orion |

[a]Exemplar systems.

Conceptually, four options are available to DoD:

- Purchase a DoD-unique satellite (using frequencies allocated to DoD, and perhaps with other DoD-unique features as well)

- Lease a DoD-unique satellite

- Purchase a commercial satellite (of an existing design, and employing frequencies allocated for commercial use)

- Lease a commercial satellite.

In fact, however, these options devolve down to purchasing a DoD-unique satellite or leasing a commercial satellite. U.S. policy assigns use of certain frequencies (such as a portion of X-band) to military satellites. Therefore, the current customers for commercial systems operating at these frequencies would be the U.S. military or U.S. allies. (This argument may hold only in the United States if commercial satellite providers move to X-band in foreign countries.) A commercially owned DoD-unique system would probably be leased for the life of the system, and the military would probably have to assume all of the capacity. We will treat this type of lease as a purchase.[19]

---

[19]The military did lease the LEASAT system, which operated in the military UHF and X-bands, from Hughes between 1984 and 1996. The United Kingdom is considering leasing an X-band payload or services hosted on a commercial satellite. In the future, commercial Ka-band systems may be designed to be tunable over the military and

Similarly, current U.S. policy assigns primary use of commercial frequencies to commercial users—meaning that military use of these frequencies must be on a "noninterference" basis.[20] In principle, the military must yield to commercial users if the two uses interfere. In addition, if the military were to purchase a commercial satellite, it would need to buy from a company with an orbital spectrum allocation for a satellite on orbit or one soon to be launched.[21]

The options we will evaluate may thus be summarized as follows:

- DoD-unique satellite

- Whole commercial satellite

- Fractional commercial satellite—a transponder or fixed bandwidth lease

- Communications service agreements.

DoD-unique satellites possess technical or operational capabilities of interest primarily to the military with no, or limited, commercial markets. We will assess DoD-owned satellites with characteristics similar to the proposed Gapfiller—a wideband satellite using military X- and Ka-band but with little jam resistance. DoD would control both the satellite payload and the bus, and could move spot beams or the satellite at will.

Whole commercial satellites currently on orbit might be leased to provide SATCOM services. (We assume that the satellites would be bought or leased on orbit so that an orbital slot assignment would be included.) In this case, the day-to-day management of satellite resources and operations would be performed by, or under the direct oversight of, DoD. The commercial satellites are presumed to utilize C-, Ku-, and Ka-band and have minimal protection.

Alternatively, DoD could lease some of the transponders on a commercial satellite or some portion of its bandwidth. The transponder/fixed bandwidth lease is similar to those leases and

---

commercial Ka frequencies, making it technically possible for them to provide capacity to both types of users.

[20]DoD (1996).

[21]The commercial C- and Ku-bands have largely been allocated already.

services offered by INTELSAT and its signatories and other commercial companies such as PanAmSat and Orion. DoD could manage the bandwidth, but the satellite bus would be operated by an outside entity (typically the private entity that has title to the satellite).

Leases come with the appropriate spectrum assignment and right to transmit in that frequency (but not necessarily the right to transmit or receive from a ground terminal without first obtaining agreement from individual nations). In addition, DoD might be able to purchase the right to switch transponders between beams at will, or the right to reposition spot beams.

Finally, DoD might procure satellite communications through service agreements with commercial providers. These vendors might own the satellites or arrange for communications on other satellites. The vendor would be responsible for procurement, network management, and allocation of satellite time to DoD users—and might be able to offer variable amounts of bandwidth as user demand changes. In the case of service agreements, DoD would control neither the satellite bus nor its payload. Service agreements may be thought of as a special class of leases, where satellites and transponders may change without notice, and DoD may not know which specific ones are being used.

These types of commercial service agreements are only now emerging. COMSAT Corporation offers a Linkway service that allows users to establish a network of terminals from a given satellite. Customers may vary the total bandwidth they use on these terminals and are billed for the capacity they actually use. Hughes Global Services has a similar concept called "DemandNet," a concept that allows users to relocate one or more terminals and still receive service. Limited capacity is available today, but these concepts are a step toward giving a user access anywhere in the world, and billing "by the bit" rather than at a flat rate. User applications under discussion for Hughes Spaceway, Lockheed Martin Astrolink, Teledesic, and other systems may make more capacity with these features available.[22]

---

[22]These systems are not yet in operation.

In evaluating the four options, we are interested not only in their ability to satisfy military demand in terms of quantity, but also along several dimensions of quality. The evaluation criteria we will use are those put forth in the Capstone Requirements Document. These criteria, which are defined in some detail in the following chapters, are as follows:

- **Capacity:** Will the option provide enough wideband throughput to meet the needs of fighting forces and their supporting infrastructure?

- **Coverage:** Will the option provide sufficient capacity in all areas needed?

- **Flexibility:** Will the option support the full range of military operations, missions, and environments?

- **Interoperability:** Will the option support the ability of all elements of the U.S. force and command structure to communicate with each other (and with allies)? Will the integration of satellites into the defense information infrastructure be transparent to users?

- **Access and control:** Will the option make the required communication services available and accessible when and where they are needed? Will DoD be able to plan, monitor, and operate the communication resources?

- **Quality of service:** Will the option provide communication channels meeting the appropriate industry standards or Mil-Specs (military specifications) for reliability, bit error rate, transmission throughput, outage responsiveness, and other appropriate factors?

- **Protection:** Will the option provide communication services that will survive attack and be robust to jamming?

In addition to the evaluations, we will consider operational concepts to mitigate shortfalls in commercial satellites or to operate combined military-commercial systems. If commercial systems are to provide future capacity, the military must use care not to introduce unintended vulnerabilities into DoD networks. That is, DoD must shift its focus from providing individual systems to the art and science of using communications for military operations.

# Chapter Four

# CAN COMMERCIAL SYSTEMS MEET MILITARY CRITERIA? CAPACITY AND COVERAGE

## CRITERIA

The next three chapters will examine the qualitative characteristics of commercial satellites and determine whether they meet military needs. We first consider the extent to which commercial systems meet the criteria of capacity and coverage. The Capstone Requirements Document defines the following characteristics as "threshold requirements" for capacity and coverage:

- Provide enough wideband capability to meet the essential needs of the deployed war-fighting forces and the out-of-CONUS (contiguous United States) requirements of DoD's supporting infrastructure

- Sustain the current ability of UHF MILSATCOM systems to provide netted service during the next decade while augmenting it with the new capabilities offered by commercial multiple-subscriber services and personal communication systems

- Provide around-the-clock coverage when and where needed on a worldwide, regional, interregional, and theater basis at all latitudes above 65 degrees south.

The following objectives were also stated:

- Support the projected growth in demand for SATCOM

- Provide narrowband MILSATCOM at data rates up to 64 Kbps from handheld terminals

- Extend coverage to all latitudes.

Finally, the CRD states that "given the commercial and military information and processing growth trends, the advanced MILSATCOM systems should be biased toward providing as much capability and capacity as affordably possible." The Capstone document defines "MILSATCOM" to include commercial capacity owned or leased by the military.

Our analysis focuses on international commercial satellite systems that provide coverage over broad areas of the globe, have systemwide standards to ensure ground terminal compatibility, and are at least partially owned by companies headquartered in the United States.[1] We will assess their capacity and whether DoD will demand a large portion of it. We will consider

- the total amount of commercial capacity in existence and estimates of expected growth,

- the subset of these amounts held by systems DoD prefers to use,

- the amount of this capacity immediately available to DoD, and the means by which more could be made available to DoD, and

- whether the supply of commercial capacity is keeping up with commercial and military demand.

## CAPACITY

Figure 4.1 compares projected global commercial capacity on C- and Ku-band with an estimate of DoD demand from the ERDB given in the Capstone document.[2]

---

[1]The theory is that U.S. companies can be relied upon to honor existing contracts to supply communications to the U.S. government. See HQ USSPACECOM (1998), p. C-8.

[2]*Ibid.*, pp. 4–8.

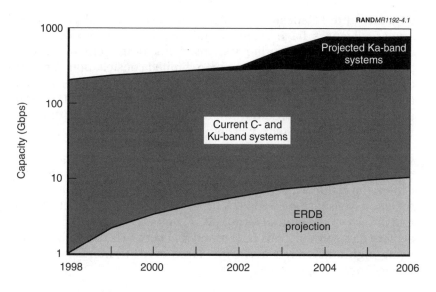

SOURCE: Euroconsult, USSPACECOM, and RAND.

**Figure 4.1—Commercial Capacity Greatly Exceeds Military Demand**

Because DoD demand is small relative to commercial supply, we plotted the graph in Figure 4.1 on a logarithmic scale so that DoD demand could be examined. (Note that the y-axis is in powers of 10.) Aggregate global C- and Ku-band capacity was calculated by summing regional SATCOM projections from Euroconsult.[3] According to our projection, even if commercial capacity did not increase, projected DoD demand would be a small portion of aggregate global capacity.

If some of the proposed Ka-band systems had been started as planned, the SATCOM market would experience a massive increase in capacity, as reflected in Figure 4.1. The potential Ka-band supply in Figure 4.1 was based upon proposals for the Spaceway, Astrolink,

---

[3]Euroconsult (1996).

GE*Star, and Teledesic systems.[4]   However, several companies granted licenses have either withdrawn their proposals or announced mergers with other systems.  Some of the remaining companies appear to be considering more limited constellation sizes and capacities than first proposed.  All of these concepts have slipped their proposed deployment dates.  For these reasons, DoD cannot assume that the Ka-band portion of Figure 4.1 will be realized or, if realized, when the systems will commence operation.

Communications capacity, especially for data and video, are often expressed in terms of data rate rather than bandwidth.  The data rate achievable within a given bandwidth depends on the waveform used, the effective isentropic radiated power (EIRP) of the satellite beam, and the size of the ground terminals.  Commercial systems can today achieve data rates of 2 bits per second for every 1 Hz of bandwidth by using high-efficiency waveforms, large earth terminals, and one channel per transponder.

Commercial providers can specialize networks to maximize efficiency for specific customers.  The military, however, must configure networks to maximize communications availability to many users.  Often, communications must be established for military users with a variety of terminal types, sizes, and ages, and with multiple channels per transponder.  These factors will tend to reduce bit-rate efficiencies.

Communications can be between large terminals, large and small terminals, or small terminals.  The Capstone document provides a "representative distribution of the requested data rates and aggregate throughput by terminal type."[5]  Combat units typically use small terminals with DoD-unique satellites.  Therefore, we assumed that DoD-owned capacity in 2008 (programmed to be approximately 3600 Mbps) will be used primarily to satisfy the demand of combat units in the field using the smallest class of terminals.

---

[4]We used the system deployment timelines, as stated in their FCC filings, of Teledesic, GE*Star, Astrolink, and Galaxy/Spaceway to determine their aggregated global capacity over a period of time (see Figure 4.3).  Where differing total capacity estimates have been publicly proposed, we have used the more modest ones.

[5]HQ USSPACECOM (1998), Appendix G.

Typically, commercial systems will be used for military communications between large fixed terminals, communications with deployed headquarters, and high data-rate broadcasts (such as video) to deployed forces. These types of traffic use large terminals (i.e., terminals delivering rates higher than 8.2 Mbps) and represent approximately 75 percent of the traffic remaining after accounting for the military capacity above. The other 25 percent uses small terminals (i.e., 8.2 Mbps or lower).

COMSAT Corporation provides communications between military users over commercial systems under the Commercial Satellite Communications Initiative (CSCI) program administered by the Defense Information Systems Agency (DISA). DISA reports CSCI data rates of 80 Mbps using 72-MHz transponders between large (11–15 meter) terminals.[6] Smaller 2.4-m terminals can transmit 20.6 Mbps at C-band or 35.4 Mbps at Ku-band through the same transponders.[7] Therefore, a bit-rate efficiency of 1.1 bps/Hz is used for 75 percent of the traffic, and a bit-rate efficiency of 0.39 bps/Hz for 25 percent of the traffic. This yields an aggregate bit-rate efficiency for commercial systems of 0.93 bps/Hz.[8]

For comparison purposes, we assumed the same terminal mix for "commercial-like" military systems (such as Gapfiller) at X- and Ka-band (i.e., 75 percent large, 25 percent small). Military terminals should operate at about the same bit-rate efficiencies as commercial contemporaries, yielding an overall bit-rate efficiency of 0.93 bps/Hz.[9] Military satellites such as Gapfiller plan to use both 500 MHz of X-band and 1000 MHz of Ka-band.[10] The Gapfiller satellite is

---

[6]Briefing by Robert Laskey, DISA CSCI director, to the Air Force Scientific Advisory Board, May 1998. In the briefing, one 72-MHz transponder supports 26 full-duplex T-1 links in Southwest Asia, for a data rate of 80 Mbps. COMSAT CSCI information gives a similar figure of 76 Mbps between large terminals. We use the higher data rate because it relates operational experience.

[7]Data from COMSAT Corporation. Receive rates can be higher if military terminals are equipped with the appropriate electronics.

[8]We assume that 50 percent of the small terminals use C-band with a 20.6/72 = 0.29 bps/Hz and that 50 percent use Ku-band at 35.4/72 = 0.5 bps/Hz. These rates assume that communications are symmetric; terminals can receive at a higher data rate.

[9]Improved military terminals may allow higher efficiencies. Commercial systems should also improve between now and 2004, when the first Gapfiller is to be launched.

[10]Assuming the military is granted Ka-band spectrum allocations.

also planned to host a "GBS Phase II–like" package onboard which will use 192 MHz of the Ka bandwidth, leaving 808 MHz.[11] At a bit-rate efficiency of 0.93 bps/Hz, Gapfiller capacity using the remaining 1308 MHz would be 1.224 Gbps. This estimate of Gapfiller capacity will be used in Chapter Seven to compare the cost per gigabit of leasing commercial capacity with that of purchasing military systems.

The global C- and Ku-band capacity shown in Figure 4.1 includes all wideband commercial systems. Many of these are high-powered systems that cover a specific region such as Western Europe or Japan and could offer useful coverage and capacity to DoD. However, we do not consider these regional systems any further in our analysis for two reasons. First, as mentioned before, DoD prefers to utilize "international" systems with at least some portion owned by U.S. companies. Second, as we will show, these systems have sufficient capacity and coverage to easily handle projected DoD demand without resorting to regional systems.

The principal international systems today are INTELSAT, New Skies, PanAmSat, Orion, and Columbia. The broad coverage of these systems allows a certain amount of flexibility in moving beams and even satellites to accommodate new customers.[12]  These characteristics are of interest to such users as DoD that need the ability to operate in any part of the world. Perhaps most important, these systems have established a working relationship with the nations covered by their beams to obtain landing rights.[13] The locations of the satellites in these systems are shown in Figure 4.2.

---

[11]The FY99 President's budget includes $404 M (then-year dollars) of spending on GBS Phases I and II. Neither the cost of providing GBS Phase II service nor its capacity is included in our estimation of Gapfiller capacity.

[12]This is not meant to imply that customers can have beams or satellites moved on request. The point is that INTELSAT, for example, can readjust the coverage of its constellation when demand increases in areas not amply served.

[13]Landing rights are the rights to receive and—when negotiated—transmit signals to and from a given nation. Past permissions do not guarantee future access; however, experience working with a given administration should make the process easier.

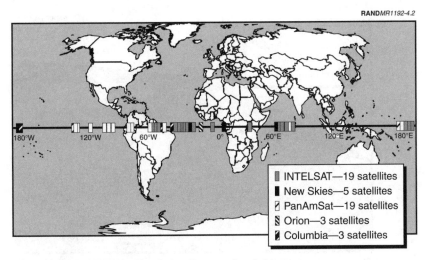

SOURCE: INTELSAT, Hughes Space Systems, Loral, Aerospace.

**Figure 4.2—Positions of International Satellite Systems**

Figure 4.3 shows capacity data and projected growth for the INTELSAT, New Skies, PanAmSat, Orion, and Columbia systems.[14] In 1996, Euroconsult estimated an annual growth rate of 4.6 percent for these systems through 2006. We applied this estimated growth through 2010 to obtain the capacity projections shown after 1999. Although only a subset of the total commercial capacity (as shown in Figure 4.1), the combined capacity of these systems exceeds projected DoD demand. Actual growth of the INTELSAT (including New Skies), PanAmSat, Orion, and Columbia systems has been faster than the Euroconsult estimate shown. The capacity of these systems, as of August 1999, is shown by the solid symbols in Figure 4.3. Since 1996 the capacity of these systems has grown by 7 percent per year.

Is commercial capacity growth keeping pace with commercial and military demand? The FCC reports that capacity leases by U.S. com-

---

[14]The INTELSAT capacity shown in Figure 4.3 includes that transferred to New Skies. The PanAmSat capacity shown includes capacity gained from the merger with the Hughes Galaxy system.

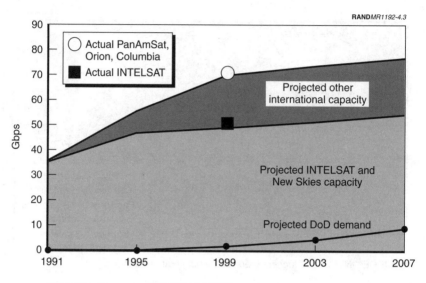

RAND*MR1192-4.3*

SOURCE: Euroconsult, USSPACECOM, and RAND.

**Figure 4.3—Projected Capacity of International Satellite Systems**

panies on international circuits increased by 13 percent annually between 1995 and 1996, and 28 percent per year between 1996 and 1997.[15] Using the Capstone demand estimates for the 2005–2010 time period, we calculate a constant compound growth rate of 15 percent.[16] In the time it takes to build and launch new satellites— perhaps one to three years into the future—we expect that further growth in commercial or military demand will encourage increases in commercial supply. Large orders from DoD will be an incentive for commercial providers to increase the amount of capacity they plan to deploy into orbit.

But how much commercial capacity is available to DoD in the near term? For needs six months to a year or so into the future, DoD could obtain additional capacity through long-term contracts offered

---

[15]See FCC Report IN 99-4, Section 43.82, *Circuit Status Data*, 1997.

[16]See also Chapter 8.

by commercial providers. The capacity would come from existing satellites that have some slack or from satellites soon to be launched that have not yet leased all of their capacity. DoD would need to buy capacity from the "ad hoc" (or spot) market to meet needs that were immediate or a few months into the future. Short-term excess capacity is provided to the ad hoc market directly by the company or through brokers. This capacity tends to be more variable, and more expensive, than that available from long-term contracts. Users can typically find some capacity unless there is a disaster or war.

## SUMMARY EVALUATION OF CAPACITY

Figure 4.4 summarizes the degree to which the four satellite alternatives meet DoD capacity needs. The key to the evaluation charts is as follows:

- **White.** The performance threshold can be achieved.

- **Gray.** Most, but not all, aspects of the performance threshold can be achieved. Non-mission-critical deficiencies will be present without additional resources.

- **Black.** The performance threshold cannot be achieved.

The threshold performance requirement for capacity is met if sufficient wideband communication capacity can be provided to fill the needs of warfighters and their supporting infrastructure. In theory, sufficient capacity could be obtained by acquiring DoD-unique satellites, whole commercial satellites, or fractional commercial satellites. In practice, however, there will be a significant lag between order and receipt of a DoD-unique satellite. The Gapfiller satellite is expected to take six years from concept to receipt on orbit. Were DoD to start today, it could not fill the projected capacity shortfall until 2004 with DoD-unique satellites alone. For this reason, the DoD-unique option is black; for future demand, it may change to white.

Commercial systems seem to have sufficient aggregate capacity today to fill projected demand, whether acquired as whole units or fractions. For the next several years, demand could be met by satel-

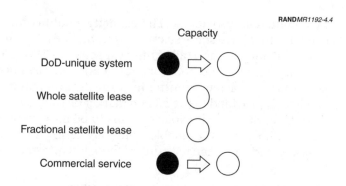

Figure 4.4—Capacity:  Wideband Capacity Can Be Provided
by a Number of Sources

lites on-orbit or under construction.  Demand further away could be
met by newly ordered units.  The whole and fractional commercial
satellite options, therefore, are white.

Eventually, enough capacity may be available through "bandwidth
on demand" or other types of communications service contracts to
satisfy military demand.  However, relatively little capacity appears
to be sold this way now.   Approximately three-quarters of
INTELSAT's capacity contracts are for terms of one or more years.[17]
PanAmSat states that 90 percent of its sales come from similar
leases.[18]  The remainder is presumably sold on the ad hoc market,
with some made available for new types of service contracts.
Therefore, the commercial service option is also black, although it
may too eventually change to white if carriers sell more capacity in
this manner.  If greatly expanded commercial capacity results from
exploitation of the Ka-band, there will be plenty of room for DoD
even if commercial rates grow.

So far we have said nothing about cost.  If one option is more expensive than another, the extra cost could inhibit achieving the required

---

[17]INTELSAT's 1998 Annual Report states that 63 percent of its analog and 89 percent
of its digital traffic are committed for 1 to 15 years.

[18]FCC Order 97-121, April 4, 1997.

capacity. We will take this issue up in Chapter Seven. Moreover, some portions of DoD's total capacity may need to be built to more demanding specifications, such as including antijamming capability, than that of commercial satellites. Capacity gained with these features will thus be much more costly than capacity without. We will explore in Chapter Nine investment alternatives and the extent to which DoD satellite budgets may constrain capacity achievable.

## COVERAGE

The coverage that commercial systems offer to DoD users is just as important as the capacity available. Commercial systems primarily cover the regions between 65 degrees south latitude and 65 degrees north latitude, as shown in Figure 4.5. The commercial world does not currently offer wideband capacity in the polar regions. The commercial market covering CONUS is relatively well endowed and competitive, so we did not explicitly consider it in our investment

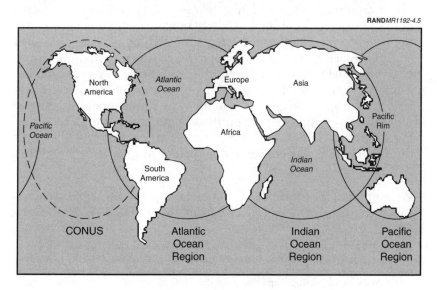

Figure 4.5—Capacity: Coverage Required for Each Strategy Evaluated

strategy either.[19]   We are left with three international regions to consider.

We are interested in the coverage that commercial systems can offer DoD over both fixed sites and deployed users.  The important fixed sites to consider are facilities in CONUS, permanent U.S. bases abroad, and Standardized Tactical Entry Points (STEPs) for the Defense Information Systems Network (DISN).  The STEPs are SATCOM terminals that receive transmissions from DoD satellites and retransmit them via leased cable circuits or other military satellites to CONUS or fixed installations abroad. DISA is considering adding commercial frequencies to the terminals at its STEPs as part of the DoD Teleport concept.

Commercial teleports and the DoD STEP sites exist within the same beam footprints of many commercial satellites.[20]  Therefore, military signals transmitted by commercial satellites could be received at either commercial teleports or appropriately equipped STEP sites. The communications could then be routed via fiber cables to CONUS or fixed installations abroad.  Military satellite communications enter the Defense Information Infrastructure (DII) by landing on a STEP site and traveling by a leased fiber cable to an entry point on a military wide-area net, and then continuing to the intended destination. Military communications traveling on commercial satellites would land at a commercial teleport and travel by similar fiber cables to the military entry point.  For the purpose of our study, therefore, receiving a signal at a STEP site, or at an adjacent commercial teleport in the same beam, was considered sufficient to relay it to the desired destination.[21]

We estimated the maximum commercial capacity available at each of the eight U.S. overseas STEPs from the INTELSAT, New Skies,

---

[19]We are chiefly interested in investment strategies for obtaining capacity in less well-endowed regions where short-term leasing costs are high.  DoD is less likely to face the same costs in obtaining capacity over the United States, where there is much available.

[20]See INTELSAT, May 1996.

[21]However, we do assume that cables cannot replace satellite delivery of communications to or within CONUS and other theaters for those needs that CRD specifically states must be filled by SATCOM.

PanAmSat, Orion, and Columbia satellites.  To estimate this capacity we identified beams covering STEP sites and added the capacity of transponders that could simultaneously use that beam.  The result of this analysis is plotted in Figure 4.6.

For deployed forces, it is important to know what beams are available from commercial systems to cover land and ocean areas. Several characteristics of modern commercial communications satellites are useful for providing connectivity to deployed forces in remote regions, including the ability to switch transponders from beams covering fixed sites to beams covering remote theaters in a matter of hours to days without moving the beams themselves. Operators could therefore shift capacity from a beam covering Europe, say, to a beam covering Saudi Arabia.  In peacetime, the capacity could be used in Europe to support day-to-day demand, and

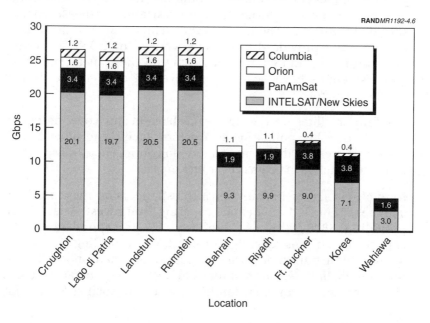

Figure 4.6—Maximum International Satellite System Capacity
Over DoD STEP Sites

could then be shifted to a beam covering Southwest Asia in the event of a contingency. This capability gives the communications planner two advantages.

- At least some of the capability needed to cover the contingency would already be within DoD's possession; hence, its availability (from a market perspective) would be assured.

- The planner would be able to purchase surge capacity in the "thick" CONUS, North Atlantic, or European markets to fill the gap left by any commercial capacity shifted to Southwest Asia.[22] Without this ability, the planner would need to purchase additional capability in the contingency-area markets, which might have scant (and hence expensive) slack.

As an example, current demand could be met in a way that would also provide options to cover Southwest and Northeast Asia, as shown in Figure 4.7. The satellites shown on the Atlantic and Indian oceans could provide day-to-day coverage for CONUS, Europe, Southwest Asia, and connectivity between these theaters. These seven satellites have beams that cover all or some portion of Southwest Asia.

The four satellites on the right each have beams covering Japan, CONUS, or other points in the Western Pacific and other beams covering Korea. In the event of a Korean contingency, transponders could be switched to beams covering Korea.

In total, 114 (36-MHz equivalent) transponders on these satellites can be connected to beams covering Southwest Asia (SWA), providing 3800 Mbps.[23] Similarly, 112 transponders can be connected to beams covering Northeast Asia, providing 3700 Mbps. Many of these transponders are in spot, zone, or hemispheric beams that contain both regions of interest for day-to-day operations (such as Europe) and a potential contingency region (such as SWA). For

---

[22]"Thick" refers to regions with many providers, ample capacity, and strong competition for customers. The FCC maintains lists of both "thick" and "thin" routes.

[23]Some of the satellites can cover all of Europe and Southwest Asia, whereas others can cover only portions. Beams considered included spot, zone, and hemispheric. Global beams were not included.

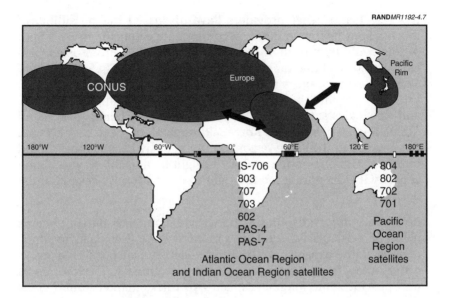

**Figure 4.7—Day-to-Day Capacity Can Help Cover Surge (1998)**

these beams, no switching is needed; the transponder would simply be used from a different location.

To illustrate a particular switching case, New Skies 703 (formerly INTELSAT 703), has a Ku-band spot covering Europe and a second covering Southwest Asia. The Ku-band transponders may be assigned to Europe during peacetime and switched to Southwest Asia during a contingency. C-band transponders, too, could be allocated to zone beams covering the Middle East. INTELSAT and New Skies satellites also have C-band coverage of Europe, Africa, and portions of Asia. The Ku-band transponders can be "cross-strapped" to the C-band beams covering Europe, Africa, and portions of Asia to provide additional flexibility.[24]

_____

[24]Cross-strapping is the combination of uplinks and downlinks using different bands—for example, a C-band uplink connected with a Ku-band downlink or the reverse.

INTELSAT and other providers have discussed the possibility of moving spot beams to better accommodate a specific user.[25] Spot beams could be moved only if they did not affect other users, or if DoD leased all of the transponders on the beam. INTELSAT and New Skies have one or more moveable spot beams on its series V, VA, VI, VII, VIIA, VIII, and the soon-to-be-launched IX satellites. INTELSAT and New Skies satellites can physically move 31 beams over Saudi Arabia and 20 beams over Korea, although all of these beams could not be used in the same region simultaneously because they would interfere with one another. Still, it might be a useful way to obtain capacity if DoD purchased the right to use and move several of the beams.

Both switching transponders between beams and moving beams might be particularly useful ways to increase DoD capacity in "thin" regions. Little capacity is currently available in such thin regions as Africa unless it is prearranged. Communications-services contracts may provide some fungible capacity in "thick" markets, but there is little incentive to make capacity available in thin markets. This might change if a wideband LEO system were deployed, because LEO systems can be built and operated to provide the same coverage around the globe (between the latitudes designed to be served).

## SUMMARY EVALUATION OF COVERAGE

The CRD defines coverage as the portion of the earth's surface to which SATCOM services can be provided. Global coverage refers to the ability to provide SATCOM services to all longitudes and latitudes of the earth. The threshold requirement for coverage is the ability to provide wideband MILSATCOM capacity when and where needed in areas north of 65 degrees south latitude. Systems are graded on this criterion in Figure 4.8.

We have shown that likely temperate-zone theaters and all DoD entry points to the terrestrial communications network are covered by commercial satellite beams with a great deal of capacity.

---

[25]In fact, COMSAT, the U.S. signatory to INTELSAT, has succeeded in having INTELSAT move beams to provide additional capacity for U.S. forces in Operation Desert Storm, the 1995 Haiti operation, and other tests and exercises.

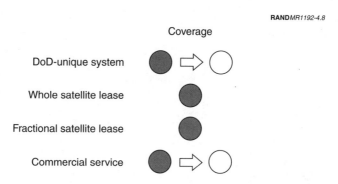

RANDMR1192-4.8

**Figure 4.8—Coverage:  Polar Regions Are Not Covered by Commercial Wideband SATCOM**

Furthermore, if arrangements are made in advance, beams can be moved by wholly leased satellites and transponders switched across beams by fractionally leased satellites, thus enhancing capacity available to DoD over high-need regions.[26] The same capabilities are of course available to DoD on its own satellites, which can also be repositioned if need be.  The commercial-service market is not yet sufficiently developed to meet coverage requirements but it may be in the future.

We are unaware of any existing systems that are able to provide wideband services to the north-polar region; hence, the options in Figure 4.8 are gray.  Commercial systems such as Iridium do provide narrowband services to this region, but it is unlikely that a significant commercial market for wideband satellite communications will emerge there.  However, depending on their chosen orbits, some future broadband LEO satellite constellations may be able to provide coverage to both polar regions.  In principle, DoD could gain access to these constellations via service agreements.  In addition, a DoD-unique satellite system could be deployed to provide wideband services to polar regions.

---

[26]DoD and commercial systems can provide open-ocean coverage with global beams. Commercial satellites can also cover much of the ocean with higher-capacity hemispheric beams, and COMSAT has tested the feasibility of "following" ships with a roving spot beam.

# Chapter Five

# MILITARY THEATER USE OF COMMERCIAL SATELLITES AND IMPLICATIONS FOR FLEXIBILITY, INTEROPERABILITY, AND ACCESS AND CONTROL

In the next two chapters, we will discuss flexibility, interoperability, access and control, quality of service, and protection of commercial systems and networks. Here, some notional operational concepts illuminate problems and possible solutions. We will grade commercial and military systems against the criteria put forth in the Capstone document.

## MILITARY THEATER USE OF COMMERCIAL SATELLITES

### Military Theater Networks

Because deployed military forces need to communicate with each other within theater and with forces and headquarters outside of theater, a theater network is established.[1] To construct a theater network, the theater commander must determine

- the people, vehicles, systems, and headquarters on the network, and their individual communications needs;

---

[1]The distinction between "within theater" and "outside of theater" may become increasingly arbitrary. The use of long-range forces from distant bases and "reachback" support tends to blur the theater boundary. We will refer to theater operations and theater networks as including those forces operating within the area of responsibility of a joint task force commander for a given contingency. The CRD defines notional major theater war (MTW) and small-scale contingency (SSC) boundaries as "2000 by 3500 km" and "1000 by 1000 km," respectively, which is useful for our analysis.

- which people, vehicles, systems, and headquarters need direct access to out-of-theater communications;

- which users need access to military systems—with the highest available levels of protection and other characteristics; and

- when it is appropriate for commercial systems to be used to augment military communications.

A simple notional theater network is defined in Figure 5.1. Some nodes represent joint command and control centers, such as Joint Task Force (JTF) headquarters or an Air Operations Center (AOC). Others represent operations centers at the major unit level, such as carrier battlegroups, brigades or division headquarters, and the headquarters of numbered Air Forces or wings. Each of the joint headquarters must be connected with the others and with each unit headquarters. The unit headquarters in turn must be connected with the fighting forces—the ships, ground units, and aircraft—that they command.

Military units have an organic capability to communicate with each other and their headquarters. The links between forward deployed

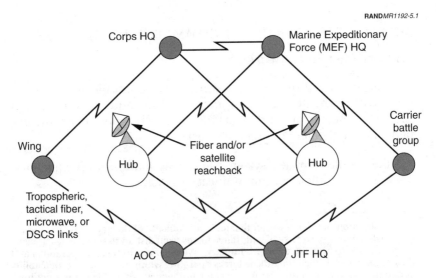

RAND*MR1192-5.1*

**Figure 5.1—Notional Military Theater Network**

ships, ground units, and aircraft and their operational headquarters are likely to use purpose-built military systems. Small ships, ground units (battalion and below), and fighter aircraft might rely mainly on line-of-site systems.[2] Large ships, ground units (brigade and above), and aircraft might also have UHF, X-band, or EHF satellite terminals that can be used for within-theater or outside-of-theater communications. Direct access to out-of-theater communications is increasing as more of these terminals are deployed.

The primary links between JTF and operational headquarters will probably use systems that can provide survivable and/or protected communications. Use of commercial terminals for these links could increase the risk of disruption such as jamming or introduce operational vulnerabilities such as direction finding. Users employing the DSCS can protect some of their communications from jamming, and MILSTAR users can add yet more protection and survivability. The difficulty, of course, is that the protected DSCS and MILSTAR capacity is small compared with the need for communications during a contingency. National users, commanders-in-chief (CINCs), and JTF headquarters will have top priority for these systems and may limit the capacity available to other users.

Commercial systems may be able to provide capacity to links to and from the theater. In addition, they might provide a backup for overtaxed, unavailable, or failing military systems.

## Military Theater Interface with Commercial Systems

A communications network can be thought of as a simple system with an origin, injection node, relay node, reception node, and destination (see Figure 5.2).[3] Information from a user enters an injection node (e.g., a satellite terminal), is routed to a relay node (e.g., a satellite), and then to a reception node (e.g., a receiving satellite terminal) that is linked to the ultimate destination. If the relay node

---

[2]These include the Digital Wideband Transmission System (DWTS), Single Channel Ground and Airborne Radio System (SINCGARS), and Joint Tactical Information Distribution System (JTIDS).

[3]Return-flow traffic simply reverses the origin and destination.

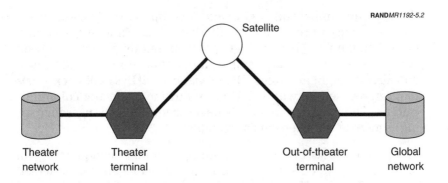

RAND*MR1192-5.2*

**Figure 5.2—A Communications Network**

is a commercial system, then there is an interface between military and commercial networks at the military user's terminal.

Networks can be constructed with few or many such interfaces between the military theater network and global commercial networks. Three examples with different numbers and locations of interfaces between the commercial and military networks are shown in Figure 5.3. In the first example, there is one interface at a central location. Other military users are connected to this central location by protected links. In this case, commercial networks provide low-cost trunk communications for the aggregated military out-of-theater traffic. In the second case, most out-of-theater communications pass through a central aggregation point, but some nodes can access commercial networks directly. In the third example, all the nodes have direct access to commercial systems. Some of the primary out-of-theater links might be commercial, with the military links serving as robust backups.

The differences in these concepts are important. The out-of-theater "hop" is vulnerable to attack, because it is close to the adversary.[4] A

---

[4]A "hop" is a satellite connection between the theater network and global networks. Physical proximity of the enemy to the theater risks direct physical attacks as well as opportunities for line-of-sight jamming.

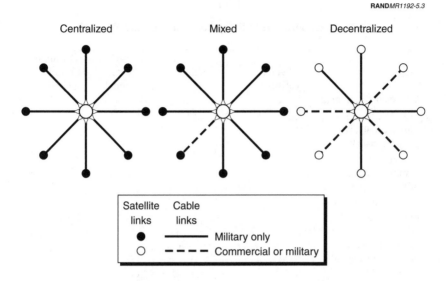

Figure 5.3—Military and Military/Commercial Networks

diverse set of routes from the military user's terminal to the global networks would make the out-of-theater communications more robust against enemy disruption. Unfortunately, there are usually few alternative routes available until the signal enters the global commercial network. Once the signal reaches an out-of-theater communications node, its final delivery is likely.

Ideally, the interface points between the military and commercial networks would be proliferated, with some located in well-protected rear areas. We constructed several notional interface concepts that differ principally in where control is passed from military to commercial systems (Table 5.1).

The first approach avoids interface difficulties with commercial systems by using purpose-built military terminals and satellites.[5] Once the signal is received by the out-of-theater military terminal (presumably located at a STEP site), it may take a second military

_____
[5]There may be difficulties interfacing military systems with each other as well.

**Table 5.1**

**Interfaces Between Military and Commercial Networks**

| Military/Commercial Interface | Theater Terminal | | Satellite Segment | | Out-of-Theater Terminal | | Global Commercial Network | |
|---|---|---|---|---|---|---|---|---|
| | Mil | Com | Mil | Com | Mil | Com | Mil | Com |
| Military delivery to global network | √ | | √ | | √ | | | √ |
| Military relay to commercial receive node | √ | | √ | | √ | √ | | √ |
| Military injection to military/commercial relay | √ | | √ | √ | √ | √ | | √ |
| Commercial injection to commercial relay | | √ | | √ | | √ | | √ |
| Parallel paths | √ | √ | √ | √ | √ | √ | | √ |

satellite hop or be routed to a commercial cable for delivery to its final destination. Military control is maintained up to the interface with the commercial cable. The military/commercial interface point can be at a protected site, but there may be only a few such sites at known locations.

The second approach uses military-unique theater terminals and satellites but could receive transmissions at either military STEP sites or commercial sites. For satellite communications, this would mean that either the satellite uses both military and commercial frequencies or that the ground sites operate dishes for military and commercial frequencies. The ability to receive the signal at a protected military site is maintained and commercial sites are added. However, there may be few such commercial sites if they need extensive modifications to receive military frequencies such as X-band.[6] The interface with commercial systems would be the same

_____

[6]This may not be a problem for military Ka-band communications if commercial terminals are built with the ability to tune over both military and commercial Ka-band

in the first approach or on the satellite downlink, which would be difficult for an enemy to disrupt.

The third, fourth, and fifth approaches are alternatives for obtaining commercial communications for the satellite segments. The third option would use DoD-unique theater terminals with the ability to connect with both commercial and military satellites (e.g., the Lightweight Multiband Satellite Terminal [LMST] or the Triband Advanced-Range Extension Terminal [START-T]). Signals from the military satellites would presumably be received at military STEP sites (unless some commercial teleports were appropriately equipped), whereas those from the commercial satellites could be received at either military or commercial sites.

The military/commercial interface would be out-of-theater (as in the first or second approaches) if a military satellite were used or in-theater if a commercial satellite were used. If an enemy were to jam the uplink for a commercial satellite, the military user could switch the terminal to a protected military link. The commercial downlink would be difficult to jam. This approach can provide an important capability to theater users as long as they have access to a number of commercial and military systems. Currently, the military is purchasing terminals capable of using commercial and military frequencies. However, if commercial "processing" satellites are deployed that use proprietary communications protocols, those satellites may not be accessible with these terminals.[7]

The fourth approach would use commercial terminals to send traffic out of the theater on commercial satellites. All military/commercial interfaces would be within the theater and hence would all be vulnerable to enemy interference. The theater user on this link

---

frequencies. If the commercial world would do so, DoD might benefit by having access through many terminals.

[7]Almost any terminal can be used with nonprocessing (or "bent-pipe") transponders so long as terminals on either side of the transmission are compatible. With some of the proposed "processing" satellites, however, terminals will need to use specific waveforms and keying protocols expected by the satellite.

would not have the option of using a military system if the commercial systems were all jammed. The fifth approach would involve the acquisition of parallel military and commercial systems. It differs from the third approach in that it includes terminals for commercial systems using proprietary protocols that multiband military terminals might not be able to access. Having both military and commercial terminals linked to the theater network might increase the interface's flexibility and robustness against disruption but the connection would add complexity.[8]

The trick would be to make the interface between the military and commercial networks simple and easily deployable, ensuring that software interfaces work properly and that any military-unique protocols are translated prior to the interface with commercial systems. The major challenge with respect to hardware will be in accommodating any unusual legacy interfaces that may still exist on the military network, and converting them to commercial standard protocols and interfaces. The selection of one approach over another hinges on determining the effects of different concepts on operational issues. Ultimately, it will come down to balancing preferences (including frequently unacknowledged ones) against expenditures.

For the out-of-theater receiving sites, both military and commercial systems would be useful. Deployable STEP sites might be sent to the area or to a large civilian teleport near the conflict. Today DoD operates STEP sites that are typically located in rear areas well away from the combat zone, yet close enough to be reached via a single satellite hop. Commercial teleports might serve many of the same functions as the STEP sites, and the proliferation of commercial teleports capable of receiving military signals would provide diverse options for theater users. Some might be co-located with military STEP sites if cross-connection is desirable. Military and commercial teleports could be equipped with a variety of links enabling access to satellite and airborne communications relays, as well as terrestrial wireline and wireless communications links.

---

[8]Instead of buying the proprietary terminal, it might be possible for DoD to obtain the rights to build a modem capable of accessing the commercial system from a military terminal.

## Operational Concepts for Transmitting Traffic Between Military Theaters and Global Commercial Networks

Within the theater, military networks will most likely use line-of-sight (LOS) wireless systems, terrestrial cables, and military satellites to connect systems, units, and headquarters. As a notional example, Figure 5.4 presents several ways to connect a military theater network in Southwest Asia to out-of-theater teleports in Europe or CONUS.

Concepts for outbound communications from a military theater network in SWA to Europe or CONUS include

- microwave or cable link to out-of-theater STEP site or commercial cable,

RAND*MR1192-5.4*

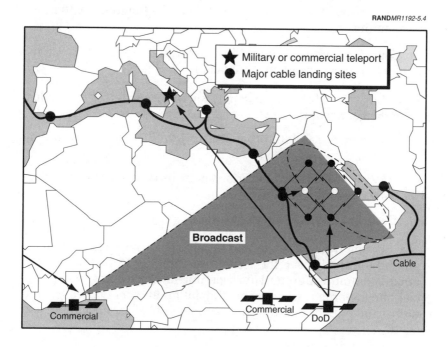

Figure 5.4—Options to Connect Theater Network with Global Networks

- satellite link through DSCS or a commercial satellite to an out-of-theater STEP site or a commercial teleport,

- link through UAV[9] airborne communications node to out-of-theater STEP site or commercial teleport, and

- direct UAV link through DSCS or commercial satellite to out-of-theater STEP or commercial teleport.

Communications outbound from the theater to Europe or CONUS may be disrupted by enemy jamming of satellite uplinks, attacks against cable receiving sites, or attacks on large satellite ground stations. Having a diverse set of alternative links may provide robustness. Military satellites, such as DSCS, also can offer protected capacity for uplinks from the theater network. As we will discuss in Chapter Six, commercial systems may be able to avoid disruption in certain cases by taking advantage of their position relative to an enemy jammer.

Concepts for communications inbound from Europe or CONUS to a military theater network in SWA include

- submarine cables connected by microwave or terrestrial cable to a theater network,

- commercial satellite broadcast to terminals within a military theater network,

- UAV airborne communications node broadcast to a theater network, and

- DSCS link with a theater network.

For inbound traffic, the submarine cable receiving sites and satellite ground terminals have the same vulnerabilities as they do for outbound communications. However, downlink jamming of commercial satellite broadcasts may be difficult for an enemy because it requires the enemy to shine the jammer into the receiving dishes.

_____

[9]UAVs are not classified as satellites and are not fixed sites. Therefore, the use of spectrum reserved for satellite traffic to fixed ground stations may be problematic. Currently, UAVs can operate on a noninterference basis with satellites and fixed ground stations. Increased UAV use in the future may necessitate specified frequency allocations. (DoD, 1996.)

Therefore, commercial satellites may be useful for transmissions into the theater even in the presence of enemy jamming attempts.

Each commercial satellite has unique properties resulting from its design, beam pattern, and location relative to a military theater. These properties might give users advantages in special situations. For instance, a satellite with 300 MHz of adjacent bandwidth available can use it to provide very-high-data-rate connections with a single user—such as an airborne sensor terminal. The degree to which the satellite is visible to an enemy affects his ability to jam, as will be discussed in Chapter Six. Commercial satellites offering these and other advantages should be identified so that DoD can utilize them.

Points the military needs to consider to ensure access to commercial systems include

- how to purchase terminal equipment that works with current commercial systems and can exploit military and commercial Ka-band,[10]

- how to encourage commercial protocol standards and the ability to upgrade terminal equipment cheaply, and

- how to make certain that the military can access many different commercial systems from the theater network.

The military is already buying ground terminals for theater networks that can exploit commercial frequencies and interoperate with commercial systems. For example, Challenge Athena and the START-T and LMST terminals can use commercial C- or Ku-bands in addition to X-band. The military will need to modify these terminals, or purchase new ones, for the Ka-band transponders on the planned Gapfiller satellite. It would be beneficial if these terminals could exploit the portions of the Ka-band designated by the FCC for commercial use, as well as those portions designated for the military. These terminals then could access the frequency bands that the proposed commercial satellites will use.

---

[10]A key capability for military Ka-band terminals is the ability to tune to both commercial and military Ka-band frequencies on the same equipment.

Of course, the ability to access the frequency band does not ensure that the satellite can be used. To employ proposed Ka-band commercial satellites, DoD should encourage industry to develop and adopt standards and negotiate DoD access to proprietary protocols. If the industry develops transmission protocol standards that are open to the military, DoD could build terminals with removable cards or plug-in equipment for different satellites. Processes to upgrade these terminals quickly and cheaply would allow the military to continue to use existing services and take advantage of new capabilities as they emerge.

The military could then use the same terminals for many different commercial systems. Locking in to one system or provider over a particular theater may make the system more vulnerable to disruption or denial, and may also result in higher prices. Military users at both fixed and deployed sites will want to have readily available options should service be unavailable on a given system. The ability to choose helps the military to obtain the best price from a group of competitors.

In the next three sections, we examine how the interface options discussed above might enhance the flexibility, interoperability, and access and control characteristics desired by the military.

## FLEXIBILITY

Flexibility is defined as the ability of MILSATCOM to support the full range of military operations, missions, and environments. It allows MILSATCOM to accommodate changing needs in changing circumstances. The Capstone document identifies no specific threshold requirements, but did identify the following objectives:

- Provide the capability to exploit new or emerging technologies readily and accommodate growing and evolving needs
- Retain/obtain sufficient frequency spectrum and orbital slots to operate military-unique and/or DoD-owned systems
- Employ multiple, diverse military and commercial frequency bands
- Maximize efficient use of limited available frequency spectrum

- Provide reliable, dependable, available, and maintainable systems, designed to facilitate ease of training and operation

- Develop and operate MILSATCOM systems under a comprehensive system safety program.

An important issue for the flexibility of military and commercial systems is frequency spectrum usage. The United States has nine "slots" for communications at 7 and 8 GHz in the X-band. These frequencies are used by the U.S. military, our North Atlantic Treaty Organization (NATO) allies, and Russia, but currently few others. The spectrum allocation lapses when the useful life of the present satellites has expired. The United States may make additional filings, "behind" the present allocations, to renew the spectrum allocation for replacement satellites. At that time, other nations wishing to use these slots for X-band communications may also file or contest the U.S. filing.

Even nations not using X-band frequencies for satellite communications may use them for aircraft and terrestrial systems and hence might object to the new U.S. filings on the grounds that they will interfere. If these nations do not themselves realize revenue for use of these satellite frequencies, they have no clear incentive to reserve them for satellite communications. When the useful lives of the present satellites have expired, these same nations may choose not to acquiesce to new systems intended to exploit these bands.

Systems utilizing C- and Ku-bands can also prepare new filings "behind" the current satellites to receive spectrum allocations after the end-of-life of the systems. These systems, too, may have trouble coordinating with the nations that will be within the visible area of the satellites. In the case of C- and Ku-bands, however, foreign nations will in most cases also be exploiting these bands for satellite communications. Therefore, even if there is a conflict over who will have the use of these bands, they will probably be clear for satellite use. Of course, even after the ITU has allocated spectrum, and the use of that spectrum has been successfully coordinated with nations in view, a nation may not allow the right to receive signals on its sovereign territory. This can delay access to foreign destinations and mean higher prices for connections in foreign nations.

The flexibility objectives can be met, at least in part, if a wide range of open architectures and various end-user terminal equipment can be supported. Military use of terminals that can exploit a variety of frequencies in order to interface with either military or commercial satellites would allow most of these objectives to be met. The ideal ground terminals for meeting this threshold requirement would be transportable, would support open architectures, and would have the ability to use a variety of frequencies. Note that the requirements for open architectures and support of legacy equipment may be somewhat contradictory. Our grading of the various systems is given in Figure 5.5.

Each satellite system received a gray grade for different reasons. Historically, DoD-unique systems have been built for a specialized set of customers with system-specific terminals and tended not to support open architectures. Moreover, for a given communications expenditure, obtaining DoD-unique satellites or whole commercial satellites would limit frequency and location diversity more than would leasing transponders on different satellites. On the other hand, spot beams on military and commercial satellites that are wholly owned or controlled by DoD could be moved quickly without having to consult other users. On the whole, these arguments led us to assign a gray grade to these two system types.

A system consisting of fractional satellites would allow more satellite location and frequency diversity. At the same time, more satellite lo-

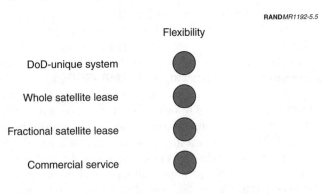

Figure 5.5—Flexibility

cations and frequencies lead to more complexity. If the military manages the network, it will increase the complexity for the military—but military management might be better for military end users, as we will discuss under "access and control." In addition, moving a fractional satellite spot beam to support fast-paced operations requires the permission of other satellite users. For these reasons, a gray grade was assigned to fractional satellite systems.

Bandwidth-on-demand would allow a great deal of flexibility in providing SATCOM to various locations on short notice. However, these commercial services may not support all DoD end-user terminal equipment. Therefore, we assigned a gray grade to these systems, but the grade may move to white if bandwidth-on-demand services mature to the point where they can support rapid deployments and fast-moving operations through military-operated terminals.

Given the desire to exploit new technologies, retain frequency spectrum and slots for military systems, and maximize access to spectrum and its efficient use, a mix of military and commercial systems is probably best. If the military owns terminals capable of exploiting commercial and military systems, it can better ensure access for a range of operations.

## INTEROPERABILITY

Interoperability is determined by how well a military user's local network interfaces with global military or commercial networks. We need to know whether commercial capacity can provide the interoperability characteristics desired from systems supporting military communications. A key question is whether communications will be ubiquitously available in a reasonably standard way, or whether they will consist of a bewildering array of unique systems and protocols.

The Capstone threshold criteria are as follows:

- CINC and JTF components (e.g., land, air, naval, mobility, combat support, and special operations forces) must be able to communicate with each other with a transparent interface between communications systems.

- Satellites (e.g., wide-area wireless communications systems) must be fully integrated into the DII, and their use must be transparent to end users.

A number of objectives are also stated (or implied):

- U.S. military forces should be able to communicate with allies, coalition partners, and other U.S. government agencies.

- Military systems must be able to interface with each other and with commercial systems.

- Military systems must provide for secure interoperability that incorporates authenticity, confidentiality, and integrity.

- All military users must have at least one communications system that allows them to exchange messages with the worldwide network (implied).

- To the maximum extent possible, legacy equipment and procedures are to be supported (implied).

Currently available military systems, commercial systems, or commercial services cannot satisfy all of the explicit or implied objectives listed above. However, there are several possible approaches that DoD could take to provide the best combination of these characteristics with today's systems and build a strategy to influence the development of new systems that will provide even greater benefits. Our focus will be on evaluating options to employ commercial systems in a way that best enhances the benefit to the military.

The threshold requirement is for interoperability between and among CINC and JTF components and for MILSATCOM to be fully integrated with the DII. The addition of terminals capable of exploiting commercial frequencies, such as STAR-T, LMST, and Challenge Athena, enhances interoperability by providing one or more additional links between and within the JTF components. In addition, the commercial systems add an alternative route into the DII through commercial teleports. A particularly difficult problem is providing interoperability with allies, coalition partners, and other U.S. government agencies.

Another difficulty accompanying the use of commercial systems is ensuring that availability, reliability, security, and protection are maintained at levels consistent with normal military operations. Building hybrid networks with both military and commercial components may introduce vulnerabilities. The problem of ensuring secure interfaces becomes particularly difficult with allies or coalition partners that may operate different and perhaps older systems.

The majority of satellites in orbit today are "bent-pipe" systems that simply retransmit the signals received. Because bent-pipe satellites do not perform onboard processing, the MILSATCOM interoperability criterion requires that various end-user terminals be interoperable. If a processing satellite or constellation is to be used, then the protocols used by that system must be considered when determining the MILSATCOM interoperability.

In this category, satellite systems all received a gray grade (see Figure 5.6) because existing end-user terminal equipment is not completely interoperable. Commercial services may move to a black grade in this category if the proposed broadband LEO constellations use protocols not interoperable with current DoD end-user equipment or other satellite systems.

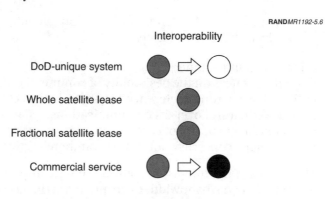

Figure 5.6—Interoperability: Ground Equipment
Interoperability Is Needed for
Bent-Pipe Satellites

## ACCESS AND CONTROL

DoD has defined assured access as "the certainty that the requisite amounts of SATCOM services are available and accessible when and where needed." Control "refers to the ability and mechanisms needed to effectively plan, monitor, operate, manage and manipulate the available SATCOM resources."

The Capstone threshold requirements are as follows:

- Provide CINC and JTF component commanders the ability to plan, allocate, and schedule accesses of their apportioned resources within fractions of hours to a few hours.

- Rapidly and dynamically configure/reconfigure resources within fractions of hours to a few hours.

Objectives include:

- Provide CINC and JTF component commanders near-real-time authorization, denial, and/or preemption of access of their apportioned resources.

- Accomplish end-to-end configuration/reconfiguration of networks within a few minutes.

It is important to understand military preferences and the effect of those preferences on the desirability of commercial alternatives. The military has a strong preference for immediate access to, and some measure of control over, its communications systems. The desire to maintain direct control of communications assets seems to lie at the heart of many concerns about using commercial systems.[11]

The origins of the interest in control seem to arise from (1) the need to manage scarce bandwidth to ensure that critical communications get through, and (2) the belief that only through direct control will the systems provide the reliability necessary. The former may be prudent in a world of communications scarcity (driven by availability or cost). The latter reflects something of the military's own culture,

---

[11]Arguments concerning the acceptability of using commercial satellites, where there is an identifiable asset, seem to arise more often than arguments concerning the use of public switched networks.

which associates the reliability and availability desired with the accountability inherent in military-operated systems. In both these instances, it is unclear to what extent control is needed over high-reliability communications networks today, and less clear how the amount of control needed will change if (or perhaps when) we enter a world of truly ubiquitous and abundant communications.

When planning the use of commercial systems, the military needs to consider several aspects of access and control:

- Commercial satellite buses will rarely, if ever, be under the control of the military. Payload control functions, such as bandwidth management, will be under control of the military for those transponders that are leased.

- "Landing rights" in foreign countries are an issue for both military and commercial systems, and may necessitate including a foreign commercial entity in the network planning.

Approval is needed to bring terminal equipment into a foreign country, to receive satellite transmissions on those terminals, and to transmit from them. Obtaining this approval, or the "landing rights," is always an issue for using satellite communications, both military and commercial. In some cases, approval has been delayed when foreign companies were not included in the communications planning.

For the military, it will be important to define a role for commercial entities—domestic and foreign—that allows them to contribute to the network planning but does not create an unintended vulnerability in network operation. As an example, commercial companies may negotiate agreements on a satellite transmission plan, the provision of terminal equipment, and connections to the public switched network in a foreign country. DoD may then want exclusive control over the operations of the military network that intends to use these assets.

The benefit-risk assessments made in the commercial marketplace are different from those made in the national security arena. The military thus has an obligation to ensure that the commercial systems can be expected to be available as promised.[12]  Commercial systems are tuned to the marketplace, where there is a fairly direct tradeoff in dollars expended to increase the reliability of a system and the economic cost of system failure.

DoD, on the other hand, may desire to decrease operational risks to systems that will be procured, perhaps leading to "requirements creep." A creep in requirements can lead to solutions that are different from those arrived at in the commercial marketplace, which is disciplined by market forces.  The danger is that the military could become less able to employ commercial communications at a time when commercial capabilities are increasing rapidly.

One way to address the problem is to impose something akin to market discipline on end users who create requirements in the first place, so that the cost of developing a purpose-built or heavily customized system is properly internalized.[13]  Conversely, it may be helpful to allow end users to obtain some amounts and types of services outside of established DoD providers.  Such a model may be similar to the Federal Telephone System procurement activity, which allows organizations to opt in or out of the system.[14]  DoD would decide how much protected capacity to provide and how much capacity to provide on unprotected DoD-unique systems.  Major commands could then obtain unprotected communications within DoD or from commercial providers, subject to some minimum constraints (e.g., for security or network interoperability).

Figure 5.7 summarizes the observations on access and control.  To meet the threshold performance requirement for access and control, CINCs and JTFs must be able to allocate capacity within their commands and plan quickly.  In addition, MILSATCOM resources must

---

[12]See DoD (1998).

[13]The Senior Warfighter Forum, or SWARF, is an effort to bring together those determining operational requirements and those allocating communications budgets. Representatives from the military services, the combatant commands, the Joint Staff, and DISA are included in the forum.

[14]The Transportation Working Capital Fund (TWCF) might be another useful analog. Transportation users can either lease commercial services or pay a comparable fee to the TWCF for military systems.  The fee is based on what a comparable commercial service would cost.

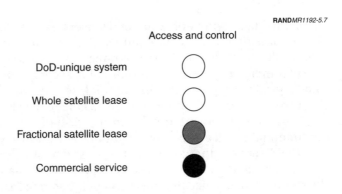

Figure 5.7—Access and Control

be rapidly and dynamically reconfigurable within a few hours. To satisfy this requirement, tools for performing network management must be accessible by MILSATCOM planners in DoD.[15] Any system that is under complete DoD control (DoD-unique satellite or whole commercial satellite) would satisfy this requirement.

A fractional satellite would meet most of the conditions of this requirement when entire transponders are under DoD control. Bandwidth on the transponder could be controlled by DoD, but spot beams cannot be moved without the permission of other users. For this reason, quick reallocation of capacity among regions may not be possible for fractional satellites, and a gray grade has been assigned to this class of systems. However, it could move to a white if the right to reassign transponders rapidly between beams throughout the satellite coverage area is obtained by DoD. These rights will need to be specified in the lease contract.

End-to-end commercial networking services providing bandwidth-on-demand may not meet the access and control threshold requirements. Bandwidth-on-demand may be desirable in terms of flexibility, but the service provider must control the use of its network assets. Military access might be degraded or denied if an ab-

---

[15]The Joint Network Management System (JNMS) might be one such tool. This tool must be able to manage commercial capacity bought or leased by DoD as well as that provided by DoD-owned and unique systems.

normally high number of commercial users try to access the system at the same time. This could happen during a disaster or war, or as the result of a deliberate hostile strategy to deny access by flooding the commercial system with requests for access. This explains the black grade assigned to commercial service in this category. However, the network provider may be able to develop processes to reconfigure the network quickly in the case of a service failure. If commercial service providers demonstrate that they can provide SATCOM services that are robust against denial-of-service attacks, the requirement for direct DoD control of network management may be reconsidered. Once providers can demonstrate "assured access," a mix of differing levels of control may give DoD the ability to enhance flexibility and retain the desired control of some systems.

Each of the various classes has strengths and weaknesses, and the ideal solution from an operational viewpoint will likely consist of a mix of DoD-unique assets that provide special capabilities and commercial systems that provide relatively inexpensive SATCOM capacity.

# Chapter Six

# DO COMMERCIAL SYSTEMS MEET THE MILITARY CRITERIA OF QUALITY OF SERVICE AND PROTECTION?

## QUALITY OF SERVICE

Quality of service refers to the ability of various systems to meet the appropriate industry standards or Mil-Spec technical standards for reliability, bit error rate, transmission throughput, outage responsiveness, and other appropriate factors. The Capstone document identified no specific threshold requirements, but did list the following objectives:

- Circuit technical performance must meet minimum performance standards required by the supported systems.

- Information must be transferred accurately and unambiguously, with minimal delays.

- Voice traffic must be intelligible and should provide voice recognition capability.

- MILSATCOM systems must be capable of degrading gracefully when operating in stressing or damaging conditions.

- MILSATCOM systems must be supported by timely, accurate space and atmospheric weather forecasts.

The first three objective criteria are categories of service quality that can be measured for both military and commercial systems. Presumably, the military can request verification from contractors that they can meet the levels of performance desired. The availability of timely, accurate weather forecasts may not be a discriminator between alternative communications systems per se, unless a particular system is more affected by weather than others (as is possibly the case with Ka-band systems). Systems that are prone to degradation from weather effects should score lower on this metric than less-affected systems. However, this metric is less a discriminator between military and commercial systems than it is between higher and lower frequency bands.[1]

U.S. military forces are increasingly dependent on reliable long-haul communications, particularly in support of deployed operations. Continuity-of-service problems can be encountered as a result of individual satellite or submarine-cable failures—whether they are caused by weather, accident, random events, or enemy action. The recent failure of PanAmSat's Galaxy IV satellite provides a case study of the far-reaching effects of the failure of a single satellite.[2] Millions of customers nationwide were affected and widely varying services including paging, point-of-sale services, and radio broadcasts were disrupted.

Eight of the ten largest paging companies in the United States leased transponder time on Galaxy IV, and the ability of these paging companies to restore services to customers varied widely. PageMart Wireless Inc., a paging company with 2.7 million customers, was unable to restore services completely until days after the failure. To restore services, Hughes had to help PageMart reposition over 2000 dishes for reception from an alternate satellite. On the other hand, SkyTel's 1 million paging customers were affected only momentarily by the satellite outage because plans were in place for continuing service after a satellite failure.[3]

The real value of commercial communications lies not in the constituent satellites, terminals, and cables, but rather in the networks

---

[1]Communications at frequencies up to 14 GHz (Ku-band) are little affected by rain fade. All communications at 20 GHz and above (Ka-band) can experience serious attenuation from atmospheric moisture.

[2]See Biddle et al. (1998).

[3]Typically, these plans include leasing "protected service," which means that the provider will move the user to a different transponder or satellite in the case of failure.

that integrate these systems. These networks can be built to provide the military needs to relay traffic to out-of-theater sites, and can provide a huge number of alternative paths to transfer this information from the sites back to CONUS. For example, the latest generation of submarine cables is typically constructed in self-healing rings that have unused or "protection" circuits designed, utilize alternative paths to provide sufficient capacity to reroute all communications on a broken main cable. In addition, alternative cables at most major telecommunications hubs have ample slack capacity to accommodate traffic routed around a failed cable.[4]

The key is to ensure that data leaving the theater have diverse available routes, so that a failure in one route does not inhibit or curtail transmission to the intended destination. It is not sufficient, for example, to ensure multiple logical paths alone—these paths might take different wires on the same cable and hence all be vulnerable to damage from a single event.[5] How do we ensure multiple physical paths between the theater forces and large commercial communications systems?

Alternatives such as microwave links, terrestrial cables, unmanned air vehicles, and satellite terminals were discussed in Chapter Five. The ability to switch between alternative commercial satellites, and between satellites and other systems, would allow communications to be routed around a disrupted link. We also discussed in Chapter Five that these alternative systems need to be able to access multiple out-of-theater interfaces with global communications networks. Having a single commercial out-of-theater site serving as the interface with a military theater switch might make communications vulnerable to a failure at the commercial site. Redundant out-of-theater

---

[4]The percentage of international cable circuits that were idle in 1997 was 37.2, compared with 3.4 percent of the international satellite circuits held by U.S. common carriers. Euroconsult estimates that between 20 and 40 percent of the usable capacity on the INTELSAT system, before the spin-off of New Skies, was idle.

[5]On Sunday, May 8, 1988, a fire in a Hinsdale, Illinois, "superoffice" is reported to have destroyed switches linking a number of central offices in the western suburbs of Chicago. As a result, wireline and wireless local, long distance, and directory services were unavailable for thousands of people over a period of up to two weeks in some cases. The concentration of so many communications paths through one physical facility was meant to increase security and flexibility in the "spokes" of the network, but resulted in one vulnerable node for the whole network. (See ACM, 1988.)

sites provide the military a choice of places to access commercial networks.

To avoid service disruptions, at least two additional steps must be taken. First, contingency plans must be in place to resolve the problem of system failures, including training personnel to deal with them. Second, it must be possible for operating personnel to rapidly reconfigure the overall communications network. The military may be able to build a robust network out of separate systems if dedicated personnel can quickly reconfigure the system. If, on the other hand, an end-to-end communications service is leased, the network management is typically handled by those operating the service and reconfiguration after a system failure is their job. To have reliable communications from an end-to-end service, it may be necessary to build a quick-restoration requirement into the contract.

The military should be able to verify whether (or not) military or commercial systems meet Mil-Spec technical standards or the appropriate industry standards for reliability, bit error rate, transmission throughput, outage responsiveness, and other factors. Communications can be degraded gracefully if the military theater network can access many different systems to carry traffic to an out-of-theater receiving site, and if diverse entry points to commercial networks can also be established. That is, the network itself may provide adequate reliability even if individual links do not. Diverse relay systems and receiving sites can be established with both military and commercial systems. All of the classes of satellite systems considered were therefore given a white grade (Figure 6.1).

There will always be a possibility that commercial—or military—systems might fail at an inopportune time and at a high military cost.[6] The cost of a system failure, whether measurable in pecuniary terms or not, may be considerably higher for DoD than for others. This would naturally affect the price that DoD (or, ultimately, Congress) is willing to pay. However, "perfection," even if attainable, is unlikely to be worth what it costs.

---

[6]These costs can include lives put at risk.

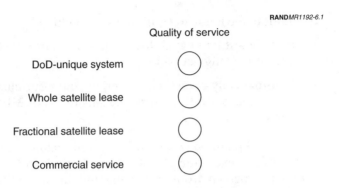

Figure 6.1—Quality of Service

It is not clear, though, that the military demands for reliable, unprotected communications are fundamentally different from those demanded by other government and nongovernment users needing high-availability systems. Commercial providers do offer some highly reliable services.

## PROTECTION

Protection refers to the ability to survive attack or defeat jamming attempts. The Capstone document lists the following threshold requirements for protected communications:

- Provide adequate survivable and antijam capacity for the National Command Authority (NCA)

- Provide MILSATCOM service for vital diplomatic, intelligence, and selected tactical users

- Provide adequate antijam capabilities for the high-priority tactical command and control and common-user networks deemed most at risk to enemy disruption

- Provide critical links or points of access that have low probability of intercept, detection, or exploitation

- Prevent unauthorized access to, and the monitoring or disclosure of, classified or other sensitive information.

In addition, the following objectives are included:

- Provide antijam capability for lower-priority tactical, strategic, and supporting networks

- Automatically detect, characterize, and neutralize offensive information operations directed against U.S. MILSATCOM systems.

Current and planned commercial communications satellites are designed to achieve optimal utilization of bandwidth at minimal cost within a benign environment. Some military communications must survive electromagnetic pulses (EMPs, typically caused by nuclear bursts); others must be able to withstand enemy jamming or defeat detection and intercept attempts.

The Capstone document states that "no current or projected commercial system" will provide the protection capabilities of the MILSTAR or planned Advanced EHF systems; calling these specialized military systems essential for the NCA and the Single Integrated Operational Plan (SIOP).[7] The Capstone document adds that "current commercial systems lack sufficient protection required to support many military requirements against deliberate disruption and exploitation."[8] The Capstone document goes on to say that commercial systems cannot protect communications from nuclear effects, hostile tactical jamming, or detection and interception. Furthermore, the Capstone document states that the Advanced EHF system will be critical to tactical users because Gapfiller "being commercial-like, will have little in the way of uniquely designed protection features."

The Capstone document is somewhat ambiguous as to precisely how much protected capacity is needed, and what the level of protection needs to be. It clearly states, however, that neither commercial systems nor the planned Gapfiller will offer the protection DoD believes is needed for the NCA or SIOP communications, jam-resistant tactical communications, or low-probability-of-intercept (LPI)/low-probability-of-detection (LPD) communications. There-

---

[7]See HQ USSPACECOM (1998), pp. 1–18.

[8]*Ibid.*, pp. 3–5.

fore, based upon the Capstone assessment, the military-unique option and the three commercial options cannot support the threshold protection requirement (Figure 6.2). The DoD-unique satellite could be designed to have protected capacity, but this would increase its costs beyond the programmed amounts. Similarly, future commercial systems might be designed with improved capabilities that offer some protection benefits.

It may be possible to mitigate the effect of enemy jamming on theater communications. In the remainder of this chapter, we will review the jamming threat to "minimally protected" military and commercial satellites, looking principally at the "tactical" jamming threat, which we will define as that posed by jammers employing moderate dish sizes and powers. Then we will discuss the use of commercial satellites in the presence of "tactical" jamming.

## VULNERABILITIES OF "MINIMALLY PROTECTED" SATELLITES TO JAMMING

Jamming is the deliberate use of electromagnetic energy to disrupt an adversary's ability to receive electromagnetic signals.[9] Jammers

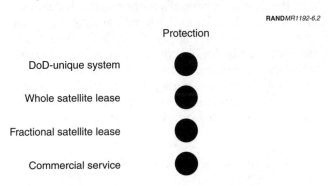

RAND*MR1192-6.2*

Protection

DoD-unique system

Whole satellite lease

Fractional satellite lease

Commercial service

**Figure 6.2—Commercial Systems Do Not Meet the Protection Criterion**

---

[9]Microwave weapons are designed to emit high-power microwaves for the purpose of causing permanent damage to receiving systems; these weapons are not considered jammers and we do not discuss them here.

can be based on aircraft, spacecraft, or ships, but our main focus is on ground-based systems.[10] Although there is a wide variety of jamming waveforms, most are "noise jammers" that have four fundamental characteristics:

- Peak EIRP[11]
- Pulse length
- Duty cycle (or pulse rate)[12]
- Bandwidth.

Peak EIRP, produced during a jamming pulse, is the most important of the four noise parameters and relates to other types of jamming as well. Smart jammers are another type that may become increasingly important. Some of these devices ("spoofers") use deception; others invoke "repeat-back" or otherwise attempt to mimic the target signal or adapt to the target signal.

Terrestrial threats to satellite communications include both uplink and downlink jamming. Downlink jamming disrupts communications from the satellite's transmitter by radiating interfering power into the ground terminal's antenna. Earth curvature and terrain obscuration makes this difficult for a ground-based jammer, unless it is close to the receiving terminal.[13] Although airborne downlink jamming can be highly effective, we assume here that friendly forces have air supremacy, so this threat can be ignored. We concentrate our analysis on the uplink jamming threat, which disrupts the communication link from a ground station's transmitter to a satellite's receiver by directing interference at the satellite.

---

[10]Shipborne jammers are a threat, but most of the issues are the same as for ground-based jamming.

[11]This measure of radiated power density (typically in watts) equals the product of actual radiated power and jammer antenna gain in the direction of the target receiver.

[12]Most noise jammers are pulsed. The duty cycle is the fraction of the time that is covered by jamming pulses; typical values are between 0.01 and 10 percent.

[13]Small, proliferated, "disposable" jammers with very low output power could be a significant problem. They might be put in place by retreating forces, for example, and operate intermittently on battery power for days or weeks.

Uplink jammers can be unsophisticated and easy to build; hence, they are the major jamming threat to satellite communications. Their major advantage over downlink jamming is that they can be effective almost anywhere within visibility of the target satellite,[14] which may cover a large part of the earth. Satellites in geosynchronous earth orbit are easy to jam for the same reasons that they are attractive for communications. First, they do not move relative to earth ground stations, so antennas need to be pointed only once.[15] Because the satellite remains stationary, a jammer's antenna needs to be adjusted only once, making it relatively easy to operate. Second, a single geosynchronous satellite can cover 42 percent of the earth's surface. Unfortunately, this wide coverage makes it possible to jam it from a significant distance.

With the exception of Motorola's Iridium satellites, all transponders on commercial communications satellites currently in orbit operate as bent-pipe repeaters. The transponder receives a signal, amplifies it, and transmits it to a ground terminal.[16] There are essentially two mechanisms by which a noise jammer degrades signal quality on such a transponder: direct reduction of the signal-to-noise (S/N) ratio,[17] and "power robbing."[18] Both sources of jamming will be included in our discussions.

Uplink jammers that have been fielded or demonstrated span a wide range of output EIRPs. If the jammer is sufficiently powerful, it need not be inside the uplink coverage footprint of the satellite, and might

---

[14]Assuming the target satellite is not equipped with special low-sidelobe antennas.

[15]A GEO satellite with a nonzero inclination tends to move slightly in the sky, but most GEO commercial satellites do not drift enough to require the ground station to actively track the satellite.

[16]In principle, a processing satellite would look for a particular waveform and perhaps some access code to verify that a legitimate user sent a received signal. If not, then that portion of the frequency band would be ignored—providing some jam resistance.

[17]S/N is the ratio of signal power to the sum of interference and normal background noise powers. This measure of signal quality determines the final demodulated error rate.

[18]"Power robbing" works on the principle that the transponder amplifies received signals in direct proportion to their transmission energy. Therefore, high-powered signals can take virtually all of the power allocated to a transponder—leaving virtually nothing to amplify other (in this case, legitimate) received signals.

even be thousands of kilometers from the theater. Powerful "strategic jammers" are examples of this type of system, with 60-foot diameter or larger antennas. Even if a strategic jammer were not located in the main uplink coverage beam, sidelobe jamming could still be possible. However, because they are fixed, they can be easily identified, located, and (when allowable under contingency rules of engagement) destroyed.

However, if the jammer is located in the same uplink coverage area as the transmitter, effective jamming of a commercial satellite can be achieved at lower power. Smaller, lower-power "tactical" and "nuisance jammers" are much harder to identify and locate (and hence to destroy).[19] Tactical jammers are medium- or high-power jammers with EIRPs that are substantially larger than those of user terminals; these would be transportable systems, with typical setup and teardown times ranging from 4 to 24 hours (shorter times might be possible for systems with smaller antennas). Mobile tactical jammers that are effective against today's commercial satellites can be built using equipment readily available from a variety of commercial suppliers. Depending on the antenna diameter, amplifier output power and bandwidth, operating frequencies, and other factors (including numbers of units to be fielded), the cost of a tactical jammer could be between $30,000 and $1 million.

Nuisance jammers are low- to medium-power systems having EIRPs on the same order of magnitude as a user terminal. These might be fully mobile, but are more likely to be transportable systems that require a short time to set up and tear down between moves. Their relatively low power makes them difficult to localize and distinguish from user terminals. A nuisance jammer could be constructed by television technicians in almost any Third World country using components that cost less than $1000. Alternatively, a typical television uplink van could be readily used as a nuisance jammer in the C- or Ku-bands.

If a commercial satellite were to be jammed today, it would be difficult to identify jamming as the source of its transmittal problem. Even where jamming could be identified as the problem, it might be

---

[19]Rules of engagement may forbid their destruction (e.g., if they are operated from within a neutral country).

impossible to isolate the jammer given the current state of the art in commercial satellite technology. Tactical and nuisance jammers could be overcome to some extent by using antennas with tight spot beams and low near-in sidelobes. Commercial satellite communications in the higher frequencies may permit tighter spot beams.

## CONCEPTS TO MITIGATE DISRUPTION OF COMMERCIAL SATELLITE COMMUNICATIONS

Although commercial satellites are not inherently designed to be protected from intentional jamming, they may be employed in a way that mitigates the disruption to communications.[20] A partial solution might be to employ some capacity on virtually every satellite over the theater—thereby complicating the task of determining which are being used by U.S. forces. This partial solution could be enhanced by having a commercial company obtain the leases in its name and sublet the capacity to DoD. An adversary would then have to jam every transponder on every satellite to cut off communications completely.[21]

A second partial solution may be to locate some theater terminals as far away from the jamming source as possible. Some "advantaged" ground stations may be able to employ a commercial GEO satellite that can be viewed at a low elevation angle above the horizon.[22] The advantage of using this satellite is that it might not be in view of an expected jamming threat (see Figure 6.3).

Suppose a friendly satellite ground station is located at Jiddah and the jamming threat is expected to be located somewhere in Iran. The ground station in Jiddah can view INTELSAT 601 with an elevation angle of approximately 6.5 degrees. However, there are no locations

---

[20]Not all of the commercial transponders in use will necessarily be vulnerable to every tactical jammer. Some transponder beams might cover some friendly units and not enemy jammers. In addition, some of the proposed Ka-band beams might be small enough to exclude tactical jammers. Strategic jammers could still enter through sidelobes.

[21]This is technically possible but might be politically difficult for adversaries.

[22]"Advantaged" refers to a terminal typically large compared with the jammer so that it has an advantage in EIRP received.

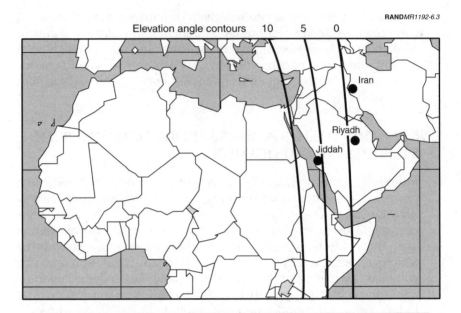

- An advantaged terminal in western Saudi Arabia can communicate with INTELSAT 601 with some periodic outages resulting from a low elevation angle (~6.5 deg)
- Jamming is impossible from western Iran resulting from line-of-sight considerations

**Figure 6.3—A Low-Elevation Satellite May Be Difficult to Jam**

inside Iran with a line of sight to INTELSAT 601. The figure shows the elevation contours for INTELSAT 601—a line of sight to the satellite cannot be achieved to the east of the 0-degree contour. Because of the low elevation, frequent service outages would be expected for communications in the C- and Ku-bands and the data throughput would be limited. The reduced data rates and intermittent service that are associated with a satellite at low elevation likely preclude it from being included in a preferred theater communications concept. However, it might be useful as a backup concept if an enemy denies other communications during a contingency.

LEO satellite constellations may be somewhat less vulnerable to up-link jamming than the more traditional GEO satellites for two reasons. First, LEO satellites move relative to the earth. To jam one, an enemy would need to buy jammers that can track the satellite. Such jamming systems are more expensive and difficult to build, which may limit the number of jammers that are employed. Also, because multiple satellites of a constellation may be in view of a ground station at any given time, multiple jammers may be required. However, the expense and difficulty of jamming all satellites over a particular theater would still be within the means of regional powers.

Because LEO satellites are moving, each satellite is low on the horizon for some time before an enemy can see it, producing a potential time window when jam-free communications are possible. Also, because of the altitude at which they orbit the earth, LEO satellites have a small footprint in comparison with GEO satellites. If the transmitter is placed at some distance away from the jamming threat, line-of-sight considerations may allow the transmitter short periods of unjammed access to satellites in the LEO constellation. By increasing the distance between the transmitter and jammer, the amount of "jam-free" access time may be extended.

We simulated satellite visibility from a point on the earth to examine the effect of standoff distance in the presence of jamming threats. To have unjammed access to the constellation, the transmitter must be able to transmit to a satellite that the jammer is not able to jam. For LEO constellations that are cross-linked, access to one unjammed satellite gives the user access to the entire constellation. We assumed that a transmitter could communicate with an LEO satellite constellation when it could see any of the satellites with a minimum elevation of 8 degrees. To jam a satellite, the jammer must also be able to see the satellite, usually with a lower minimum elevation—which we assumed to be 5 degrees.[23]

As our first example, we simulated a small LEO constellation at the low end of orbits proposed for LEO systems.[24] We calculated visibil-

---

[23]The jammer has the advantage that it may jam the satellite with relatively "dirty" signals.

[24]Our purpose is to look at the geometric visibility of LEO constellations. We used 66 satellites in orbits of 764 km in altitude with six planes inclined at 86.4 degrees. This is

ity for users needing 8 degrees of elevation (the apparent satellite angle above the ground as measured from the users' site) to communicate with a satellite, and users needing 25 degrees. The jammer should be able to disrupt communications for satellites with an elevation of 5 degrees as seen from his site. By varying the distance between the transmitter and jammer, we find the relationship between daily constellation access time and standoff distance (see Figure 6.4).

Access time to the satellite for the constellation grows roughly linearly with the distance between the transmitter and the jamming threat. When the transmitter is located more than 2000 km from the

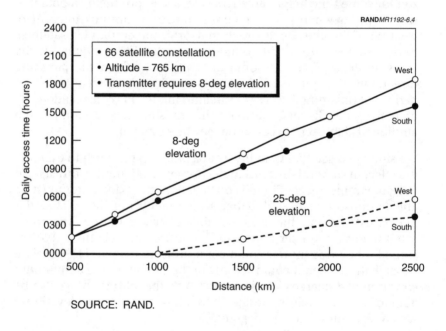

Figure 6.4—Transmitter Access to LEO Constellation Increases Linearly with Distance from Jammer

similar to the Iridium system, but because no features of the satellite are considered (i.e., transmitter characteristics, satellite antenna characteristics, satellite processing details, etc.), our example does not describe jamming threats to Iridium.

jammer, it is possible to access the satellite constellation for more than 12 hours per day with an 8-deg elevation angle. However, the access to the constellation is intermittent, as shown by the white bars for a two-hour span in Figure 6.5.

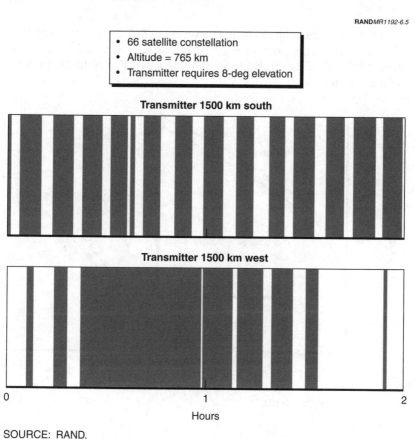

Figure 6.5—Jam-Free Access Period

For our second example, we assessed a larger constellation with a higher altitude,[25] both for users requiring 8 degrees of elevation to see the satellite and users needing 25 degrees. The jammer is assumed able to disrupt communications on satellites with an elevation of 5 degrees.

Once again, the relationship between access time and transmitter distance from the jammer is roughly linear (Figure 6.6), and access time to the satellite constellation will be intermittent in the presence of a jamming threat.

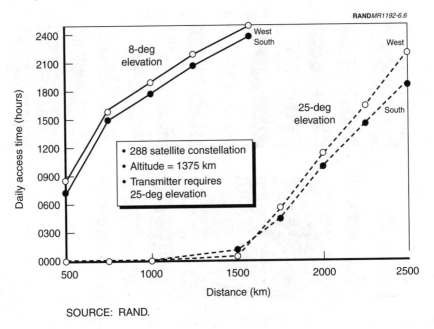

Figure 6.6—Visibility of LEO System in Teledesic Orbit

---

[25]We used 288 satellites at 1375 km in altitude arrayed in a constellation of 12 planes inclined at 84.7 degrees. This is similar to that of the satellite constellation currently proposed by Teledesic. Once again, we are not assessing the jamming threat to Teledesic, because we consider no aspect of the satellite itself. We examine only periods of nonmutual visibility from disparate ground sites to a system in a similar orbit.

In some circumstances, a satellite might be accessed directly from an airborne platform—for example, if the airborne platform were itself a communications node. In this case, it may be passing communications it receives from other terminals on to satellites that cannot be seen by theater tactical jammers. Another case of interest may be if the airborne node is a sensor platform. The Capstone document reports that airborne sensors may be the source of much of the data to be passed out of the theater. Currently, these data may be downlinked to a ground terminal, which may then retransmit the data by cable or satellite out of the theater. Alternatively, a UAV may transmit directly to a satellite, in which case finding a commercial satellite not in view of a jammer may be of great interest.[26]

An airborne platform at high altitude (e.g., 60,000 feet) is located above weather effects such as rain and clouds that adversely affect communications in the Ku- and Ka-bands. Because of the lack of ground clutter, an airborne platform may be able to communicate with commercial satellites at 0-deg elevation. A jammer directly underneath the aircraft, at the same longitude and latitude, will typically need at least 2 degrees of satellite elevation over the horizon to even see the satellite because of ground clutter. (More elevation might be needed to be certain of jamming the satellite, but we will assume for this calculation that merely seeing the satellite is sufficient.) To determine the difference in elevation between jammer and aircraft as a function of aircraft standoff from the jammer, we again modeled an illustrative example. In this model, the jammer is located in western Iran at ground level and the transmitter is located on an airborne platform at 60,000 feet (see Figure 6.7).

In Figure 6.8 we plot the celestial arc visible to the airborne platform but invisible to a ground-based jammer with a minimum elevation

---

[26]These links may not necessarily operate at the same data rates. For some systems, the link to the ground site may operate at a higher data rate than that to the satellite link.

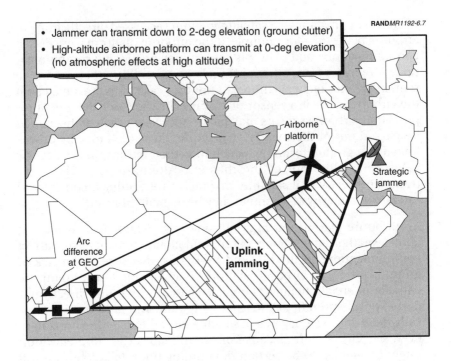

Figure 6.7—Airborne Platform Using a Low-Elevation Satellite
to Overcome Strategic Jamming

angle of 2 degrees, for various horizontal distances from the jammer. For example, when the jammer is located directly below the airborne transmitter, the airborne platform could see a satellite in 2.3 degrees of celestial arc to the west of the aircraft that the jammer could not see. This arc difference can be significant. If the jammer were located in western Iran, the airborne platform could transmit without the jammer's interference to INTELSAT 601 when the jammer is located directly underneath the platform.[27] If the airborne platform is located to the west of the jammer, then the platform can see a larger arc along the geosynchronous orbit.[28] The arc can also be ex-

_____

[27]Of course, this requires that the satellite, INTELSAT 601 in this case, have a spot beam that covers the area overflown by the airborne platform.

[28]The upper line in Figure 6.8.

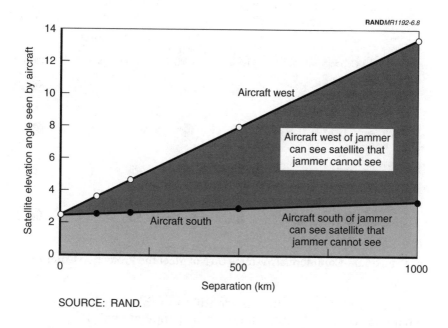

SOURCE: RAND.

**Figure 6.8—Maximum Elevation of Satellite Visible to Aircraft but Invisible
to Ground-Based Jammer**

panded somewhat when the platform is located to the south of the
jammer, but the increase is not as pronounced.

## EVALUATION SYNTHESIS

We next synthesize our evaluations of criteria from Chapters Four
through Six (Figure 6.9). Looking at the results as a whole, it does not
appear that any single class of system can satisfy the requirements
defined in the Capstone document. Use of commercial systems is
essential if DoD is to achieve the communication throughput rates
desired in future contingencies for any budget level not greatly ex-
ceeding the current one. However, commercial options require the
military to accept lower levels of access and control than it would like
and much lower levels of inherent protection than it believes necessary. These and other commercial-system shortfalls relative to
Capstone criteria can be mitigated through the implementation of

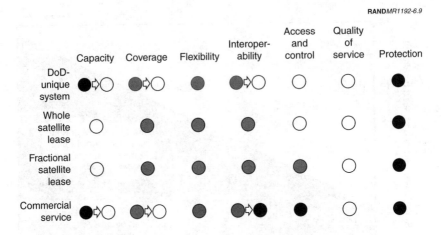

Figure 6.9—Commercial Systems Do Not Meet the Protection Criterion

various operational concepts identified in this and the preceding chapters. Ultimately, the solution from an operational viewpoint will likely consist of a mix of DoD-unique assets that provide special capabilities and commercial systems that provide relatively inexpensive SATCOM capacity.

# Chapter Seven

# PRICES OF COMMERCIAL AND MILITARY CAPACITY

In this chapter, we estimate the cost of buying communications satellites and leasing transponders or bandwidth services. To understand more about the communications market, we will discuss some of the major providers and products in the commercial market, then estimate prices for DoD to lease capacity. Finally, we will estimate the price to DoD of purchasing a "commercial-like" satellite.

## COMMERCIAL MARKET PROVIDERS AND PRODUCTS

From 1964 until 1988, INTELSAT, the only satellite system providing international communications services, sold leases or communications services through companies that were signatories to its charter. Leases could be long-term commitments or short ad hoc arrangements. Signatories to the INTELSAT agreement were also the shareholders of INTELSAT and often the only companies authorized to sell INTELSAT services in their country. COMSAT Corporation, the U.S. signatory to INTELSAT, arranges all leases for communication between users in the United States and INTELSAT satellites. Foreign companies then arrange "companion" leases between their nations and the satellite for the other half of the communications link.

Until recently, COMSAT was the only authorized provider of INTELSAT services in the United States, and so was categorized as a "dominant carrier" by the FCC. This obliged COMSAT to publish a tariff with the FCC listing the prices of the various services. These tariffs had to be cost-justified and could not provide COMSAT with a rate of return that was considered to be excessive. The tariffs could be changed with 14-day prior notice, and new ones could be offered to respond to the needs of specific customers.

In 1996, COMSAT was given a waiver (except for video service) for cost-justifying its rates for the space segment. Voice service was waived partly because little international voice traffic remained on the satellites. In 1998, the cost justification was lifted entirely for all space-segment services with the exception of "thin" routes—those with little competition. For the other routes, no cost justification is needed and tariffs can be added or changed with one-day notice, but the new tariffs now need to remain in effect for a minimum of 30 days. The tariffs represent a standard price for a given service, and FCC regulations direct that these tariffs reflect the prices paid by customers to COMSAT.[1] As of September 1999, INTELSAT will sell directly to users within the United States. Direct sales of INTELSAT capacity are allowed in more than 65 countries worldwide.

In 1988, PanAmSat became the first competitor to INTELSAT by providing satellite services between North America and several Latin American countries. Now, INTELSAT,[2] New Skies, PanAmSat, Orion, and Columbia Communications all sell satellite communications services over much of the globe. PanAmSat sells long-term leases or service commitments directly to users. PanAmSat has not been classified as a common carrier by the FCC in part because it does not have a standard price for its long-term leases but negotiates prices with each prospective user.

INTELSAT leases the bulk of its capacity for commitments of 1 to 15 years, and most of these are for 15 years.[3] The prices are established in a tariff filing, which states prices for varying types of service, length of terms, data rate or bandwidth, power, coverage, and several other

---

[1]From conversations with the FCC. However, RSI, a division of COMSAT, was able to lease transponders from COMSAT at a substantial discount for the CSCI contract to DISA. (FCC Order 97-315, 1997.)

[2]INTELSAT has transferred five satellites to New Skies, a subsidiary that will be spun off to INTELSAT shareholders. New Skies is intended to be a first step toward privatizing the INTELSAT network; hence, its operations should more closely resemble other commercial companies.

[3]1998 INTELSAT Annual Report, p. 41.

factors.[4] PanAmSat also leases most of its capacity on a long-term basis, although, as mentioned above, it does not offer a standard price.[5] Historically, satellite communications providers like to make long-term leases because such leases guarantee that a portion of a satellite will be filled and hence provide a stable income.

Long-term contracts can be preemptible or nonpreemptible. Preemptible service can be taken away from a user, after an agreed-to notification period, by another user with a higher priority. Notification time can vary from instantaneous (no warning) to hours, days, or weeks depending on what has been negotiated. Conditions for preemption can also vary greatly, from "business need" (meaning that a higher-priority user now has need for capacity previously relinquished), to failures on other satellites or cable systems serving higher-priority users. Significant discounts are given to buyers of preemptible capacity, with the size of the discount increasing as the notice period shortens or conditions for preemption become less stringent.

Alternatively, users can buy capacity with various levels of assurance. These begin with nonpreemptible capacity, meaning that no other user has a higher-priority claim on this capacity and it will not be taken away if capacity becomes scarce or if other users experience system failures. A higher level of assurance is "protected" capacity, meaning protection against loss of service (and not to be confused with any antijam capability). Protected service can be purchased to ensure availability even if the satellite experiences some failure—say of the transponder or a portion of the satellite power. In such a case, even nonpreemptible users may lose service (depending on the terms of their leases), but the provider would have enough preemptible capacity onboard the satellite to serve the protected users. An even higher level of protection would be protection against the failure of the entire satellite. In this case, capacity on another satellite would be given to the protected users to maintain service. Each progressive increase in protection comes at a higher premium.

---

[4]FCC Order 99-236, September 1999; COMSAT Corporation (1998).

[5]FCC Order 97-121, 1997. At that time, PanAmSat leased less than 10 percent of its capacity in the ad hoc market.

Capacity not sold long term is often sold on the ad hoc market. Ad hoc leases are of short duration—e.g., for days or weeks. They use some of the capacity that the satellite owners have not been able to lease on long-term contracts, or capacity that the owners are keeping in reserve in case of a satellite failure. In the event a failure occurs, some of the ad hoc leases are preempted and the capacity shifted to other users who have purchased guaranteed capacity.

Ad hoc capacity is sold directly by the satellite owners and can be resold by customers who have excess. Often, capacity is resold through brokers who find new customers and negotiate terms. Ad hoc capacity is usually much more expensive per unit of time than long-term commitments. Terms for ad hoc capacity can vary enormously by length of time, preemption conditions, size, power, and so on. For service to U.S. customers, COMSAT has established tariffs for many variations of the above parameters. PanAmSat has issued a "rate card" for ad hoc capacity, but uses it as the "basis to begin negotiations" and reserves the right to change the terms at any time without notice.

Who leases long-term capacity commitments? Who leases on the ad hoc market? Users with established long-term needs will typically lease with long-term commitments. Examples include telecommunications companies needing satellites for voice and data transmissions, and cable companies needing to broadcast video to local cable television networks. Recently, direct-to-home broadcasters have emerged that have specific needs for satellites in precise locations and very high power to service small user dishes. These users need carriers to provide a consistent, reliable service for long periods of time and hence are willing to engage in long-term contracts. These users will typically lease more capacity in advance of projected needs and sell excess capacity in the ad hoc market. (We assess the benefits of obtaining capacity in advance of projected need in Chapter Nine.)

Users who need to obtain capacity quickly, without advance warning, often use ad hoc capacity leases with commitment terms of less than one year. Ad hoc uses include satellite news-gathering that may require a large amount of capacity for short periods to cover disasters or special news events. Even large users occasionally buy ad hoc capacity to cover unexpected demand peaks. Interestingly, ad hoc capacity is not the dominant means for large media concerns to cover wars. Instead, they increase their leases of one-year service commitments in those cases (such as Kosovo) where they see tensions building.[6]

How does the market react to "atypical" conditions—for example, when a calamity or war occurs? It depends, of course, on the communications users involved, the scale of the calamity or war, and the size of the communications market in the area affected. Large users of communications, such as news organizations, increase their long-term leased capacity to handle the expected increase in demand.[7] Additional capacity may be leased on the ad hoc market by smaller users or users needing to cover a demand peak.[8] In "thin" markets, i.e., those served by few systems, sudden demand surges can use up all available capacity. To some extent, this happened in the Middle East market during the Gulf War in 1990 through 1991.[9] In "thick" markets capacity may remain available throughout the conflict for terms of one year or more.[10]

In addition, other users holding leases on these systems may be willing to relinquish them for a price. Sometimes leaseholders ask the satellite owner to "sublet" some unneeded capacity on their behalf. The sublet terms might mirror the original contract and extend for the time remaining on that contract. Capacity "resellers" or "brokers" might also be asked to sublet capacity. The sublet may be for a shorter time (e.g., days or weeks), and may be subject to preemption if it is needed by the original leaseholder. Even if leaseholders offer no capacity for sublet, they may be willing to relinquish some capacity if DoD were to offer a sufficiently high price.

---

[6]Discussions with Hughes Global Services, Inc.

[7]DoD also acquired additional commercial capacity in Bosnia and Kosovo ahead of actual operations.

[8]Conversations with Hughes Global Services, Inc.

[9]Conversations with COMSAT Corporation, May 1997. News organizations and DoD used most of the capacity available in the region at that time. At least one additional beam was moved over the region to cover demand.

[10]As an example, 198 MHz of bandwidth were available for immediate lease over Kosovo (with a one-year duration) on April 30, according to one broker. The transponders providing this bandwidth varied greatly in terms of other areas covered, power level, and price. On August 4, after the conflict, 180 MHz were available over Kosovo from the same broker.

In theory, ad hoc prices listed in the FCC tariffs could change during a contingency as long as the required notice is given. In practice, we have seen no evidence of fluctuations in tariff prices over the last three years as a result of small-scale contingencies. However, ad hoc capacity sold by carriers or brokers other than INTELSAT, and not listed in tariffs, may rise during a contingency.[11]

## COMMERCIAL MARKET PRICES

We calculated the price of 1 Gbps of service (a convenient measure) for one year using four illustrative capacity contracts from the FCC Tariff 3 for COMSAT World Systems. The results are shown in Table 7.1.[12]

The first two are ad hoc contracts, and the second two are long-term commitments. From the ad hoc market, we examined a series of one-week and three-month leases, and then assessed long-term leases of one and ten years.[13] Prices for leasing whole, standard-power C- and Ku-band transponders are from the 1998 COMSAT Tariff 3 for the U.S. half-channel, and a factor of 1.377 times the U.S.

### Table 7.1

#### Lower Prices in Exchange for Longer-Term Commitments

| Acquisition Method | Price (Gbps-Year) | Duration |
|---|---|---|
| Transponder lease | $274 million | One week |
| Transponder lease | $154 million | Three months |
| Transponder lease | $77 million | One year |
| Transponder lease | $58 million | Ten years |
| Satellite purchase | $48 million | Ten years |

---

[11]Hughes Global Services saw no evident rise in ad hoc prices during the war in Kosovo. Prices may, of course, rise in future wars.

[12]The COMSAT tariffs represent standard prices for services (this does not guarantee the capacity availability).

[13]The one-week lease rate quoted was for preemptible capacity.

price was used for the foreign half-channel.[14] We will use these contracts to determine under what conditions it makes sense to purchase capacity ahead of demand.

The shortest contingencies in our database lasted roughly one week, so this is probably the minimum time for which the U.S. military would lease. The median length of contingencies was three months, so contracts of this length should also be considered. One- and ten-year contracts bound the prices for long-term commitments. Also shown is the estimated annuity cost of 1 Gbps of capacity obtained by purchasing a "commercial-like satellite" similar to the proposed Gapfiller system.

Our cost calculations consider the effects of price trends and the opportunity cost of money.[15] The varied nature of telecommunications applications and lack of transaction documentation make generalizations about price trends difficult. We used a declining annual price trend of 4 percent, and discounted future expenditures at a 3.6 percent annual rate.[16] This real rate is based on that of equivalent-maturity U.S. Treasury bonds and does not necessarily reflect the social opportunity cost of employing federal funds.

---

[14]The foreign half-channel prices on CSCI and other government contracts in Europe, Japan, Korea, and Australia (Organization for Economic Cooperation and Development [OECD] nations) are roughly equal to the U.S. price. Hence, these regions and ocean regions (where there is no foreign half) were assigned a value of one. All other regions were assigned the same premium paid in Saudi Arabia—2.5 times the U.S. price. Capstone projections were used to determine the fraction of demand per region.

[15]Some analyses do not consider price trends or the opportunity cost of money. This has the effect of assuming constant real prices and a zero time-value of money. The first assumption is inconsistent with telecommunications history; the second is unrealistic.

[16]The FCC ordered COMSAT to reduce tariffs 4 percent annually on "thin routes" as a condition for reclassification as a nondominant carrier. The FCC determined that "thick routes" are competitive, and reports real-price declines of 8 percent for international switched telecommunications (see FCC, 1998). Recently, COMSAT has announced price cuts of 25 percent for certain transponders, and FCC Order 99-236 allows U.S. customers to access INTELSAT satellites directly, without paying the full COMSAT markup. This ruling is expected to cut satellite transponder costs between 10.7 and 35.2 percent. A 3.6 percent discount rate is in accordance with Office of Management and Budget (OMB) Circular A-94, dated November 1992.

Some interesting steady-state price comparisons can be made with Table 7.1. A particularly large differential exists between the one-year and one-week contracts: A one-year transponder lease costs between 26 and 30 percent (depending on the contract terms) as much as purchasing the equivalent capacity through 52 one-week leases.[17] The one-year contract price is 28 percent of the one-week contract price we use here, and 50 percent of the three-month price. The ten-year contract is 75 percent of the one-year price. The annuity price for purchasing 1 Gbps of capacity is 62 percent of the price of the one-year lease, and 83 percent of the price of the ten-year lease.

We used the Gapfiller program as the exemplar system for purchasing DoD-unique capacity. The Gapfiller satellite is planned to carry two-way X- and Ka-band transponders as well as the GBS Phase II, which is a Ka-band transponder for broadcasting communications to forces in the theater. Cost data for Gapfiller came from the 2000/2001 Budget Estimates for the price of research, development, test, and engineering and purchase of three satellites.[18] The deflated total research, development, test and evaluation (RDT&E), acquisition, and launch expense programmed for three Gapfiller satellites is $1.219 billion. Launch costs are accounted for in the evolved expendable launch vehicle (EELV) program element, and are programmed to be $193 million in FY98 dollars.

Gapfiller capacity and cost can be calculated either including or excluding the GBS capacity. Three GBS II packages are now carried on the UHF Follow-On (UFO) satellite, and three more are planned to be carried on the Gapfiller satellites. Of the total programmed GBS Phase II cost of $455.9 million, $264.8 million is planned to be spent after completion of all UFO launches. Therefore, some of this cost should be attributed to the three GBS packages planned for Gapfiller.[19] However, no information is available to estimate this

---

[17]Large price differentials can also be seen for purchasing small amounts of bandwidth rather than full transponders. For example, the price for 72 MHz of bandwidth is roughly half as much as leasing 72 MHz of bandwidth with 720 100-kHz leases.

[18]See Department of the Air Force (1999).

[19]The GBS Phase II equipment is hosted on the UFO 8, 9, and 10. The "Joint Wideband Gapfiller Concept of Operations" draft (see HQ USSPACECOM, 1998) states

breakdown, so we decided to count neither the cost of the GBS package nor the capacity it adds to the Gapfiller system in our estimates.

The U.S. government does not insure launches, but it should include estimated losses of launchers and satellites into expected costs. A rate of 17 percent of the insured value is fairly typical for launch and operations.[20] The total cost of each satellite and launch vehicle, including a pro-rata share of RDT&E, was $380 million, or $445 million including the premium for expected loss. This is equivalent in value to ten annuity payments of $53.8 million each at a discount rate of 3.6 percent. Wholly owned satellites also incur an annual operating charge. We estimated $5 million per year for satellite tracking, telemetry, and control.[21] Together, the total is $58.8 million per year per 1.224 Gbps or $48 million per 1 Gbps-year (see Table 7.1).

We have shown that the annuity price of buying satellites yielding 1 Gbps of capacity is cheaper—in the steady state—than leasing the same capacity. Why do we not conclude that we should always buy satellites rather than lease capacity? There are several additional factors to consider.

As noted in the flexibility assessment of Chapter Five, it is important to have the ability to access both commercial and military frequencies. Under current FCC practices, the military can use commercial frequencies only on a noninterference basis, and hence cannot be the primary user of these frequencies on DoD-owned systems. Therefore, DoD needs to lease commercial frequencies from a system owned by a commercial entity.

More important, we have treated demand as though it were known with certainty. In fact, the demand in the face of contingencies is highly variable—as we shall show in the next chapter. We have yet to show whether it is better to make long-term capacity commitments ahead of time to cover contingencies or to lease surge capacity for

---

that it will carry "a GBS package, compatible with and comparable to GBS on UFO" and use the same theater injection and payload control network.

[20] This amount was reported for insuring the latest INTELSAT satellites—802 through 806. From FCC DA 97-958, DA 97-2036, DA 97-2037, DA 98-1134, and DA 98-418.

[21] The per-satellite cost of running the Air Force Satellite Control Network is roughly $5 million once operations, support, personnel, construction, and procurement are taken into account.

short periods of time on the ad hoc market. Purchasing is a form of long-term commitment, so we must address the effect of demand variance on cost before we answer this question.

Finally, the emerging requirements database and the Capstone estimates project a large growth in communications demand—hardly a steady-state environment. How DoD times its increases in capacity will be important, and strategies to match the supply of capacity and demand need to be evaluated. Practical differences between leasing capacity and buying satellites will become clear. For example, buying satellites will require more up-front spending than leasing. In addition, although leased capacity may be available shortly after the decision is made, a DoD-unique satellite will not be. Several years may be required for the contractor to build the satellite once it is ordered, and that delay will affect the costs borne by DoD. The effect of these dynamics on expected costs will be treated in the next two chapters.

Before leaving our comparisons of commercial and military systems, we will make two additional observations. As mentioned in Chapter Two, the military must have an allocated orbital slot to operate a satellite. Currently, DoD has nine orbital slots for 500 MHz of X-band communications[22] and must place DoD-owned satellites in these slots. Does placement of a satellite in one of these slots "use the slot up"? That is, does the operation of one satellite in a slot make it impossible to operate another satellite in that slot? If so, then we must estimate the value of the slot consumed by this use and add it to the estimated cost of military communications. If not, then we can ignore the estimated slot value.

We believe that the answer is no: The operation of a satellite in an orbital slot does not make it impossible to operate another satellite in the same slot. As mentioned in Chapter Three, several commercial companies operate multiple satellites in the same slot. Careful coordination of beam location and frequency use is needed to prevent interference, but it can be done. Several military satellites can occupy the same slot, each pointing its beams over different areas

---

[22]DoD has filed for the use of 1000 MHz of Ka-band communications in the same slots.

within their view. Also, one or more of the satellites in the nine slots could be spares, and would be moved in case another satellite fails.

As a final note, our comparison of Gapfiller and commercial system data rates assumes technologies currently in hand. Technologies will emerge that will allow higher capacities both for Gapfiller and for the commercial systems.[23] It is extremely important when making comparisons between these systems to ensure that they are compared on an "equal technology availability date" basis.

----

[23]Although we estimated an average of 0.93 bps/Hz for military use of commercial systems, there are commercial modems with large terminals that provide 2 bps/Hz.

# Chapter Eight

# VARIANCE IN MILITARY COMMUNICATIONS DEMAND

In this chapter, we discuss the variance of future military communications demand. After providing some context, we show how uncertainties arising from the timing of contingencies can result in substantial variations from a smooth demand projection.

## DEMAND PROJECTIONS AND NATIONAL SECURITY STRATEGY

In 1997, United States Space Command (USSPACECOM) estimated total current DoD demand to be roughly 2 Gbps,[1] which includes day-to-day support activities and daily military operations. Day-to-day operations include the NCA and DISN backbone, the Diplomatic Telephone Service, and capacity for strategic forces, indications and warning, and intelligence agencies. Daily military operations include routine patrols, tactical intelligence, training and exercises, and "current crisis" (such as Southern Watch and Northern Watch). Additional capacity would be needed to support a major theater war.

Current national security strategy is to be prepared to fight and win two overlapping major theater wars. How is this strategy supported by the ERDB and Capstone demand projections? The ERDB demand projections are given in Figure 8.1. If DoD had 9300 Mbps of ca-

---

[1]From a briefing by Commander Baccioco to the SWARF, August 1997, Peterson Air Force Base, Colorado, and early Capstone analyses. An estimate of 1 Gbps was attributed to the ERDB in the Capstone document. We use the higher estimate because actual DoD-owned capacity was roughly 850 Mbps in 1997, and 958 MHz of bandwidth was leased commercially through CSCI.

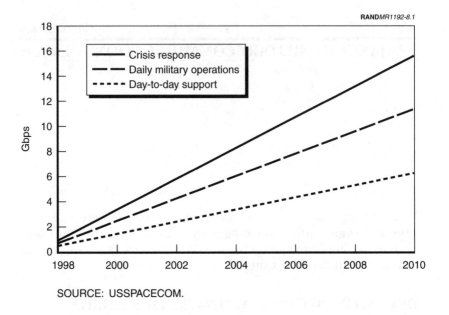

SOURCE: USSPACECOM.

**Figure 8.1—ERDB Projections of MILSATCOM Capacity Demand**

pacity in 2008, for example, it would have enough to satisfy total day-to-day demand. If a contingency were to occur in 2008, DoD could take capacity away from "daily military operations" to satisfy contingency needs, obtain enough additional capacity to satisfy the contingency, or do some of each.[2] If a major theater war were to occur, DoD could take all 4000 Mbps of capacity needed away from daily military operations, or obtain an additional 4000 Mbps of capacity and support both the MTW and daily military operations, or some mix of the two. If a second MTW were to occur, then one MTW could be supported by using all the capacity allocated in peacetime to daily military operations, and the second MTW could be supported by obtaining an additional 4000 Mbps of capacity.

---

[2]Communications planners at USSPACECOM state that the "daily support activities" must be supported in peace or war. Therefore, we assume that this capacity cannot be taken away to provide some of the capacity needed to support contingency operations.

DoD could acquire the capacity needed to support peacetime needs and the "two overlapping MTW strategies" by having only the capacity needed in peacetime and a "surge" DoD capacity when a contingency occurred. In this approach, DoD could meet contingency needs but would be trying to obtain capacity at the last moment, when other users (such as the news media) might also be trying to buy more capacity. Some capacity may be available for periods of a year or more, but much of the capacity may need to be purchased on the ad hoc market, which tends to be for short periods of time and more expensive than capacity purchased for longer periods. DoD might have to purchase expensive capacity in order to satisfy contingency demands.

On the other hand, DoD could acquire more capacity than needed for day-to-day communications for periods of a year or more. This spare capacity could then be used to satisfy contingency demand as it arises. However, the spare capacity would need to be paid for all of the time—even when no contingencies occurred.

Which approach is better: Waiting until a contingency occurs to obtain capacity, or acquiring some capacity ahead of time even though it must be paid for even without a contingency? Before we can answer that question, we need to know more about contingency demand. We need to know how often contingencies typically occur, and how much capacity is needed to satisfy contingency demand. Planning would be simple if contingencies occurred in a smooth, constant fashion. Unfortunately, they do not. The remainder of this chapter will treat the variance in expected contingency demand.

## ASSESSING THE EFFECTS OF CONTINGENCIES ON DEMAND

The Capstone projections tell us what we might expect, *on average,* in a given future year. They do not (and cannot) tell us how demand might vary from week to week within a year, or how much capacity might be required in a "bad" year—i.e., a year in which many contingencies arise. The contingency demand component is a function of the size, type, location, and duration of military operations. To adequately consider the effect of these factors on communications, we looked at the historical pattern of contingencies and assessed how future surge demand might vary if recent historical trends applied.

The Capstone document describes the crisis response demand in terms of two MTWs or four small-scale contingencies (SSCs). Major theater wars occur rarely, and past experience may not help us to predict when the next will occur. However, the United States has been involved in many SSCs throughout the 1990s, and we can calculate the number ongoing at any one time and the variance. To do so, we reviewed the SSCs to which U.S. forces were committed from 1990 through 1997. Contingencies of one week or less and involving 100 or fewer people were eliminated from the analysis. Humanitarian operations within the United States or with troops already deployed for other reasons were also eliminated.

The remaining contingencies were then grouped according to size, with those involving fewer than 5000 troops classified as small SSCs, those involving 5000 to 15,000 troops as medium SSCs, and those with 15,000 to 45,000 troops as large. The occurrence and duration of the contingencies in each of these categories may be seen in Figure 8.2.

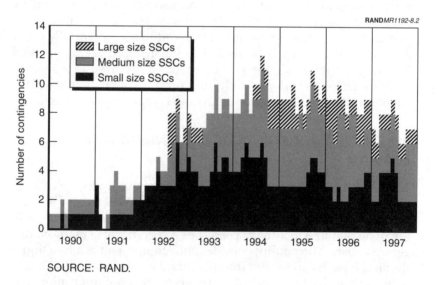

SOURCE: RAND.

Figure 8.2—Historical Small-Scale Contingencies

Approximately ten new contingencies, on average, occurred each year, with a mean value of 6.5 contingencies ongoing at any one time. Of these 6.5 contingencies, on average 2.9 were small, 2.7 were medium, and 0.9 were large. On average, the large contingencies involved approximately 25,000 troops; the medium contingencies, 10,000 troops; and the small contingencies, 2500 troops. We weighted these operations, based on size, by assigning a value of 1.0 to the large contingencies, 0.4 to medium contingencies, and 0.1 to small contingencies. This scheme resulted in a weighted average value of 2.3 contingencies ongoing at any one time (see Figure 8.3). The Capstone document estimates the communications demand of a "generic" SSC as 1690 Mbps in 2008. We estimate that our large SSC is the equivalent of Capstone's generic SSC; planning for 2.3 of them suggests a 2008 surge demand of 3900 Mbps—if it is an average year.[3]

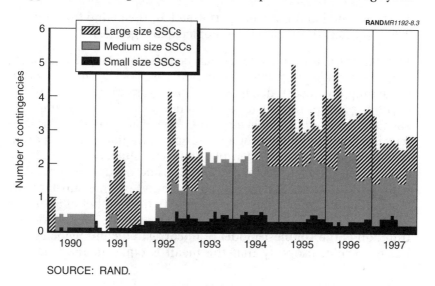

SOURCE: RAND.

**Figure 8.3—Historical Small-Scale Contingencies Weighted by Size**

---

[3]Our estimate is based on two facts: the Capstone scenario description of a small-scale contingency large in scale and scope, and the total demand close to 50 percent of the demand expected in an MTW. Both facts suggest a "large" SSC.

Simply setting aside 3900 Mbps for SSCs would not ensure sufficient capacity in a given year—the contingencies might "bunch up" over a short time. To determine the effect of uncertain contingency timing on communication demand, we built a model to simulate the future occurrence of SSCs assuming frequencies and duration similar to those in the historical data. This mathematical model is a Monte Carlo simulation with a one-week time-step and a secular increase of 15 percent per year. The growth rate was calculated by fitting a curve to the day-to-day demand in 1997 and that projected in 2008. Within each six-month period, five SSCs may begin randomly, with no more than five SSCs simultaneously. Each of the five SSCs continues for three months and has a 60 percent probability of ending and a 40 percent probability of being extended for a second three months independent of the other SSCs.[4]

Ten thousand iterations were performed for the period from 1998 to 2008. A typical simulation run is depicted in Figure 8.4.

To summarize, the United States has been involved in 2.3 equivalent large SSCs on average over the period 1990–1997. Our model for simulating the occurrence of SSCs allows from zero to five SSCs to occur at random times with a frequency such that on average 2.3 SSCs are ongoing at any given time. The average is thus straightforwardly derived from experience. The assumption that SSCs occur randomly, although intuitively plausible, is slightly more complex to derive from the historical record. If SSCs do occur randomly (i.e., with no correlation of events), over a period of time substantially longer than the typical duration of an SSC, the number of SSCs as a function of time should form a Poisson distribution.[5] A Poisson distribution is the simplest model of uncorrelated contingencies and has the useful property that the mean is equal to the variance.

---

[4]The five-simultaneous-SSC limit, median duration of three months, and 40 percent probability of extending reflect data observed in actual contingencies 1990–1997. A limit of five simultaneous SSCs may seem less than the "two nearly simultaneous MTW" strategy, but each of the five generic Capstone SSCs need nearly half of the communications capacity of an MTW. Hence, in total, five SSCs need more capacity than two MTWs.

[5]We find that a fit of the historical record to a Poisson distribution cannot be excluded with high confidence (greater than 89 percent). This does not prove that SSCs are highly random, but does demonstrate that with the limited amount of data available randomness cannot be excluded with high statistical confidence.

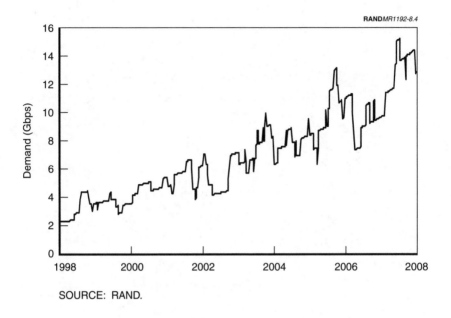

SOURCE: RAND.

**Figure 8.4—Typical Simulated Demand Over a Ten-Year Period**

Because we set the mean of our model equal to the mean of the observed data, the variance of our model will match the variance of the data. We use this model to investigate alternative investment strategies.

# Chapter Nine

# INVESTMENT STRATEGY ALTERNATIVES

In this chapter, we estimate the expected costs of alternative investment strategies for DoD to obtain communications capacity over the next decade. Specifically, we seek answers to the following questions:

- Is it more economical for DoD to make long-term capacity commitments in advance of contingencies[1] or to use short-term leases to satisfy contingency demand?[2]

- How do various buy-only and lease-only strategies compare in cost?

- How do we guide the decisionmaker in procuring commercial capacity in an uncertain future?

In answering the first question, we want to know if it ever makes sense to purchase more capacity than we need to satisfy day-to-day demand. Our analyses will consider alternative investment decisions with some degree of abstraction. To answer the second question, we consider dynamic factors affecting DoD investment decisions, including criteria for making decisions to obtain capacity such as the timing of capacity acquisition relative to demand and the effect of

---

[1]As discussed in Chapter Three, buying DoD-unique satellites and leasing commercial satellites are the two readily available options. For the remainder of this report we will assume that purchases relate to DoD-unique systems and leases to commercial systems unless otherwise noted.

[2]Short-term leases here refer to leases of less than one year in duration. As examples, we use rates quoted for one-week and three-month leases.

order-receipt lags on cost. For the third question our intent is to present strategies that rely on what is known (or knowable) about prevailing conditions or what can be surmised about the short term (a year or less).

We build our strategies upon the following assumptions:

- Current military demand is known

- Current commercial offerings and prices are known

- Specific contingency occurrence and demand are unknown, but we expect the general historical patterns to continue.

## DETERMINING AN ECONOMICAL BALANCE BETWEEN BUYING AND SHORT-TERM LEASING

A key question is whether DoD might spend less by (1) buying capacity in advance of surges or (2) making ad hoc acquisitions of just enough capacity to meet demand as it surges. In the second approach, shortfalls would be satisfied by entering markets to obtain contracts that just meet demand—and hence might typically be short-term or small in quantity. We will use the demand simulation discussed in the previous chapter to assess whether DoD would ever want to obtain capacity for contingencies in advance given expected prices and demand growth. We assume that long-term capacity can be leased or bought at the prices given in Chapter Seven, and evaluate the cost of the following conservative baseline strategy:

- Determine current DoD demand

- Obtain enough capacity to satisfy current demand in increments of 1 Gbps.[3] It is assumed to take one year for this capacity to become available for use.

- Engage in one-year lease contracts to satisfy the current need for communications while waiting for the purchased capacity to become available.

---

[3]In this section, we used purchases, although in principle leases of one or ten years would have yielded a similar result. We will examine the difference between satellite purchases and capacity leases in the next sections.

- Use one-week lease contracts to satisfy contingency needs as they emerge

- At the end of the current year, repeat the process for the next year.

This baseline strategy conforms to a simple rule of thumb—buy only what is currently needed. These guidelines are intentionally conservative, in that little or no slack capacity is held (assuming day-to-day demand does not decrease during the year). Because contingencies in this strategy are served with week-by-week contracts, no slack will be carried when the contingencies are concluded. A typical demand simulation is shown in Figure 9.1.

The short-term variation in demand visible in the graph is the result of SSCs occurring at unpredictable times. Each run varies from the next, of course, in the timing, frequency, and duration of contingencies, and these variations strongly influence the resulting variations

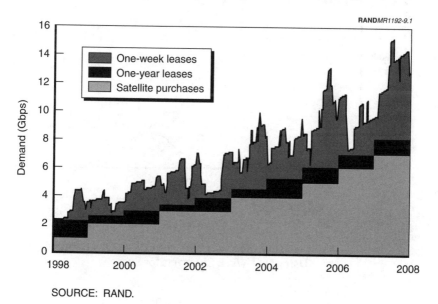

SOURCE: RAND.

Figure 9.1—Typical Monte Carlo Demand Simulation

in expected cost—most contingency demands occur over too short a period to warrant purchase of a satellite and are met instead with high-cost one-week leases. Long-term increases in demand have much less effect on our cost results because they can be met with lower-cost satellite purchases.[4]

We ran 10,000 simulations of this strategy to estimate the expected costs of this strategy according to the rules set out above. The total cost of each simulation run was calculated, and Figure 9.2 gives the frequency distribution of these costs. By inspection one can see that, over ten years, the baseline strategy could be expected to cost about $6.7 billion. Depending on the level of contingency activity over the

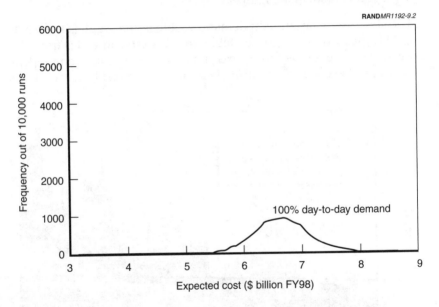

Figure 9.2—Expected Cost of Buying Only for Day-to-Day Demand and
Using One-Week Contracts for Surge

---

[4]Long-term decreases in demand could similarly be met by retiring satellites or not renewing leases.

period, expenditures could be as low as $5.5 billion or as high as $7.9 billion. The frequency diagram is not symmetric; it has a longer tail on the right side, because contingency operations can increase expenditures above those for day-to-day demand but will never decrease them.

The baseline strategy has the virtue that DoD never buys SATCOM capacity that it does not use. The downside of these guidelines is that DoD buys large quantities of expensive one-week leases. What if DoD tried to cut down on the number of these leases by purchasing more capacity up front? The left-most curve in Figure 9.3 shows the distribution of costs from a strategy in which DoD purchases the equivalent of 20 percent more capacity than necessary to meet its day-to-day needs each year.

This strategy results in total spending of about $5.4 billion—or $1.3 billion lower than the baseline. (For symmetry, the right-most curve

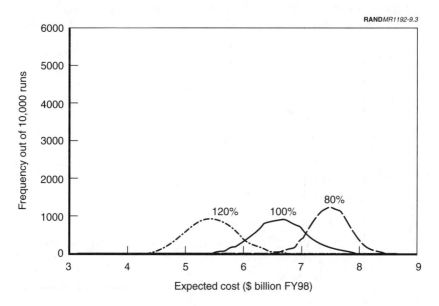

Figure 9.3—Effect of Buying Slightly Less or More Than Day-to-Day Demand and Using One-Week Contracts for Surge

gives the distribution of costs for buying 20 percent less than necessary to meet day-to-day needs and making up the difference with one-year leases.) Clearly, the baseline is too conservative in its long-term contracting, given the assumption that demand is increasing and the price ratio of long-term to spot-market purchases is low. The frequent use of expensive one-week leases costs significantly more than any savings resulting from avoiding periods of slack capacity.

To find even more economical acquisition strategies, we continued to increase the amount of capacity purchased over the baseline amount until we detected an increase, rather than a reduction, in expected expenditures. At that point, the cost of carrying excess capacity begins to outweigh the benefits of lower long-term prices. Figure 9.4 shows the original baseline result and five alternatives of purchasing an increasingly greater percentage of day-to-day demand.

In the example illustrated in Figure 9.4, the cumulative cost to the government is lowest when DoD obtains the equivalent of between

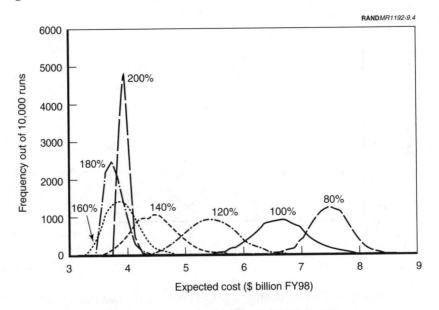

**Figure 9.4—Effect of Buying Increasing Multiples of Day-to-Day Demand and Using One-Week Contracts for Surge**

60 and 80 percent more capacity than it needs to satisfy day-to-day demand. Purchasing more than this amount increases the expected expenditure over the ten-year period. We also note that the cost of erring on the high side is lower than the cost of erring on the low side. Expected cost declines $0.7 billion by purchasing 180 percent rather than 140 percent of day-to-day need, and by $2.2 billion by purchasing 140 percent rather than 100 percent of day-to-day need. DoD should avoid purchasing less than 140 percent, because expected costs increase more rapidly below this point in the curve.

We can see from the narrowing of the distribution in Figure 9.4 that the uncertainty in cost to DoD decreases as DoD makes more long-term purchases. This is because purchasing additional long-term capacity decreases the contingency-driven demand variation (and thus the cost variation) from one run to the next. Thus, the greater the reliance on long-term acquisition, the more certain DoD will be about its actual level of expenditure.

Of course, this result is sensitive to the price differential between short-term and long-term purchases. As discussed in Chapter Seven, three-month contracts are roughly half of the price of one-week contracts on a per-month basis. How does the expected expenditure change if we were to use three-month ad hoc contracts rather than one-week contracts? To examine the effect of cheaper ad hoc contracts on expected expenditures, we reran our analysis using three-month ad hoc contracts (Figure 9.5). As can be seen, purchasing approximately 60 percent more capacity than needed to meet day-to-day demand resulted in the lowest expected cost.

What if DoD could obtain one-year contracts, at the one-year tariff, on the ad hoc market? We would expect that DoD would not want to purchase more capacity than needed to satisfy day-to-day demand if it could obtain a one-year lease as soon as a contingency occurred. In fact, our analysis supports this expectation. DoD would not save money to purchase extra capacity if one-year leases were available.

We observed earlier that, in the current market, the price differential of year-long and week-long contracts is substantial. However, the price differential between one-year and ten-year contracts and between ten-year leases and outright purchases is significantly less.

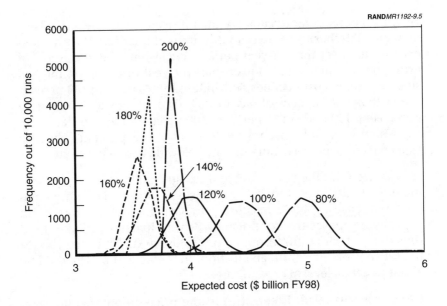

Figure 9.5—Effect of Buying Multiples of Day-to-Day Demand
and Using Three-Month Contracts for Surge

Results of the foregoing analysis, although illuminating, are limited by assumptions we made about communications demand and prices. We assumed that the number of future contingencies would resemble our experience over the past decade, and that demand for both day-to-day capacity and capacity for contingencies would grow at an annual rate of 15 percent. Also, we assumed that the price ratio of each of the lease contracts to the purchase option would remain the same. A reasonable question to ask is, "How would the results of this analysis change if we changed some or all of these important assumptions?"

We could answer this question the same way we did before—by assuming different combinations of expected demand and prices, generating new curves, and finding the capacity amount that minimized expected cost over some time period. Alternatively, we could relate expected cost to prices and demand probability, and solve the problem analytically. This approach would yield a curve like the one in Figure 9.6.

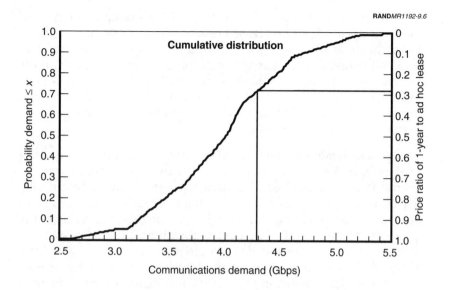

Figure 9.6—Finding the Capacity That Minimizes Expected Cost

A general, graphical method can be used by DoD communications planners to find the capacity they should buy in advance to minimize expenditures.[5] The first step is for DoD communications planners to draw a graph of the cumulative distribution of communications demand expected over the next year.[6] The X-axis index is communications demand, in gigabits per second, and the Y-axis index (on the left-hand side) is probability. The curve then is the cumulative probability that demand will be less than or equal to a given amount of capacity. We derived this curve from the historical data used in our earlier simulation.[7] The Y-axis on the right-hand side is the price ratio of one-year leases to ad hoc leases.

---

[5]See Appendix C for the derivation of this technique.

[6]It could include the following year if planning must be completed two years in advance to fit within the budget cycle.

[7]We have drawn this particular graph for one year only (1999 in this case) rather than the ten-year period shown earlier.

The graph is used in the following way:  If, for example, the ratio of one-year leases to ad hoc leases is 0.28, a horizontal line is drawn from the 0.28 point on the right-hand axis to the cumulative probability distribution curve.  By dropping a vertical line from the intersection with the curve to the horizontal axis, the planner obtains the long-term capacity level that minimizes expected cost—just under 4.3 Gbps in this case, or 165 percent of the 2.6-Gbps day-to-day demand expected in 1999.

This graphical method can be used to answer the questions we posed at the beginning of Chapter Nine.  If expected demand were to change relative to the historical data, we could draw a new cumulative probability distribution curve.  If the price ratio of one-year to ad hoc leases were to change, we would enter the right-hand side of the graph at the new ratio.  The communications quantity associated with the lowest expected cost would then be found as before.  Although the technique was illustrated using simulated data from a model of global demand, in practice it could be used in each regional market separately, to reflect both the uncertainty and the price ratio unique to each market.

## DYNAMIC INVESTMENT STRATEGIES

Our comparison of investment strategies up to this point suggests that

- it may be more economical to make long-term commitments and "waste" some capacity than to underbuy and make up the shortage with short-term service contracts

- what is most economical is determined by the expected demand and the long-/short-term price ratio.

Recall that these findings apply when resources can be brought to bear as needed to ensure the most economical outcome.  In practice, however, an acquisition strategy must be planned that delivers capacity as it is expected to be needed.  Some flexibility should be built into this strategy so that it can respond to variations in demand and conform to a budget profile.  We have said nothing so far about the timing of purchases relative to demand increases or the effect of a delay between the order and receipt of a satellite on expected costs.[8]

In this chapter, then, we seek to fulfill two objectives:

- Determine how well a variety of strategies meet demand and the effect of order-receipt lags and other factors on expected cost

- Determine which of these strategies is the most economical in terms of having the lowest present cost.

## APPROACH

To explore the implications of these factors, we constructed the demand curve shown in Figure 9.7. Demand was projected by applying a 15 percent annual growth rate to current day-to-day demand and

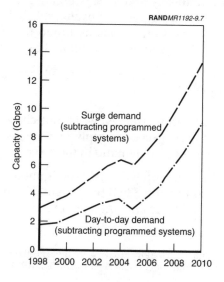

**Figure 9.7—Can an "Optimal" Strategy Be Implemented Given Estimated Demand and Budget?**

---

[8]That is, if DoD can obtain a lease in the same year the decision is made but must wait five years to obtain a satellite from a DoD-unique development program, that will affect our expected cost.

subtracting from that the capacity expected from present and pro-grammed DoD satellite systems.[9]

"Excess demand" must be satisfied by purchasing a DoD-unique satellite or leasing commercial capacity. We chose to use a demand growth rate that is consistent with USSPACECOM estimates and Capstone projections; however, our analytic approach works with different rates.

As noted in Chapter Seven, DoD plans to meet half of the communi-cations demand expected in two major theater wars (the surge de-mand) by obtaining additional capacity and the other half by reassigning capacity from daily military operations. Surge demand for two MTWs (or four SSCs) is projected to be approximately 140 percent of the day-to-day demand in 2008. We therefore drew the upper curve at 140 percent of day-to-day demand. To meet the increase in both day-to-day and surge demand, we examined four acquisition and lease strategies and evaluated their costs.[10] Each time a purchase was chosen, it was assumed to be placed in the geographic region most needing the additional capacity.[11]

We developed four investment strategies to fill the demand depicted on the above graph:

- A "leading" acquisition strategy, which purchases a satellite ev-ery year that demand exceeds supply. This leads to some excess of supply above expected demand in most years.

- A "following" acquisition strategy, which purchases no satellite until it can be completely filled by current demand.[12] One-year leases are obtained to fill any unmet demand.

---

[9]For military-owned capacity, USSPACECOM gave the estimates for the capacity ex-pected to be available from DSCS, DSCS SLEP, MILSTAR MDR, Gapfiller, and Advanced EHF.

[10]Recall that at the price ratios used in this analysis, obtaining capacity to satisfy all of the expected surge demand is cheaper than obtaining leases on the ad hoc market.

[11]The satellite could be placed in the slots allocated to DoD in whatever way makes most sense—e.g., several in one slot, spread evenly among them, and so forth.

[12]The implicit assumption is that each region has identical demand or that capacity can be traded between adjacent regions.

- A lease strategy in which ten-year leases are engaged to fill exactly the entire demand.

- A lease strategy in which one-year leases exactly fill demand.

Costs of the satellites purchased and leased under these strategies are as given in Chapter Seven.

## Baseline Results

We first examine the leading acquisition strategy (Figure 9.8). Under this strategy, DoD purchases three satellites in the first year (1998) to meet existing day-to-day and surge demand. For the moment, we assume that the new satellites arrive instantaneously, with no lag between purchase and receipt. (Time lags of three and five years will be examined next.) Additional satellites are purchased in subsequent years as demand again rises above the initial amount of capacity held. Once the Gapfiller satellites begin deployment in 2005, excess DoD demand dips momentarily and no new satellites are pur-

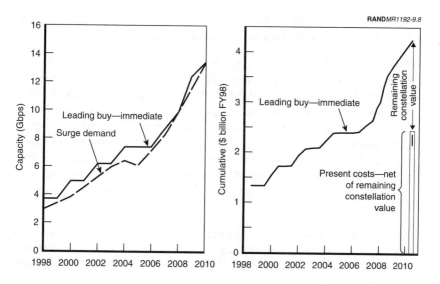

Figure 9.8—"Leading" Acquisition Strategy Exceeds Near-Term Budgets

chased until 2007. Excess demand then accelerates, and additional satellites are purchased every year through 2010. The cumulative total expenditure for this strategy from 1998 through 2010 (as shown in the right-hand graph) is roughly $4.3 billion in FY98 dollars.

To determine this strategy's net present value, we must account for the value of the remaining life of the satellites operating at the end of the planning period (as shown by the gap in 2010 between the cumulative cost curve and the vertical line). In fact, those satellites purchased late in our strategy should have almost their entire useful life remaining. To value this remaining life, we assumed that each satellite would have a productive life of ten years.[13] The cost of operating those satellites every year of their life after 2010 was compared with the cost of leasing the equivalent capacity with one-year terms. We considered the money saved by operating the constellation until the last satellite ended its useful life to be the remaining constellation value. This value was then subtracted from the cumulative cost of the leading acquisition strategy, giving the cost (net of remaining constellation value) as shown by the vertical line in Figure 9.8.

Next, we examine the "following" acquisition strategy, and compared its performance with the leading strategy (Figure 9.9).

The "following" acquisition strategy waits to purchase a satellite until its capacity could be completely filled. In the meantime, one-year leases provide the capacity needed. In our exemplar calculations, the capacity of two satellites could be filled in 1998 and hence are purchased. As communications demand continues to grow, more satellites are purchased when their capacity can be completely filled. The "following" acquisition strategy purchases one satellite less, during the years we examined, than does the "leading" strategy, and has a lower cumulative present value cost than the "leading" strategy.

We note that the two satellites purchased in the first year of the strategy could not cover all of the Capstone regions. Before acquisition of the satellites, DoD would have leased capacity. Presumably, these

---

[13]This is the design lifetime of INTELSAT 7A and 8 series (INTELSAT 7 was 10.9 years). See Martin, 1996.

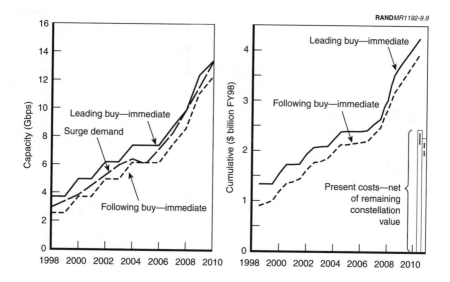

Figure 9.9—"Following" Acquisition Strategy Also Exceeds Near-Term
Budgets but Ultimate Discounted Cost Is Less

leases could be planned in such a way that the capacity no longer
needed in a region could be switched to other regions or terminated.
The one-year and ten-year leasing strategies acquire exactly the
amount of capacity needed in each year (Figure 9.10).[14]

In general, less cash is expended each year by leasing than using either acquisition strategy. The leasing option appears to cost more
than the acquisition strategies when we account for the remaining
value of leases and acquired satellites.[15] However, less total cash
outlay is required over the 12-year planning horizon.[16]

---

[14]The ten-year lease strategy actually carries a slight excess in 2004 when Gapfiller is
deployed, but the amount is small.

[15]The ten-year lease strategy results in some leases still in effect after 2010, but including this effect changes the total cost of the ten-year lease strategy only slightly.

[16]However, unlike the acquisition strategy, the ten-year lease strategy commits DoD
to a continued stream of payments beyond the planning horizon, unless the leases can
be sold.

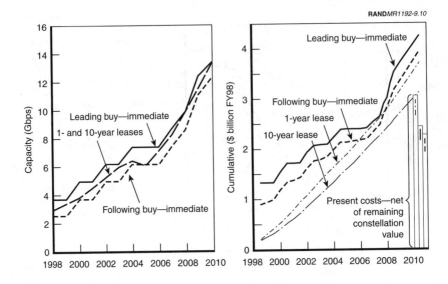

RAND*MR1192-9.10*

**Figure 9.10—One- and Ten-Year Leasing Strategies
Are Closer to the Budget**

On the right-hand side of Figure 9.11 is a *cumulative* estimate of the DoD budget available to fund alternative wideband investment strategies.

To arrive at this estimate, we took the DoD estimate of the remaining money planned for wideband systems after Gapfiller is completed,[17] and we added current estimates of the money spent every year by DoD to lease commercial capacity.[18]

During an actual contingency, Congress might allocate additional funds to obtain the communications needed for the operation.[19] However, because we do not know when these contingencies will occur, and hence when any additional money will be available to ob

---

[17]From President's Budget for FY1999.

[18]We received estimates of $200 million or higher from USSPACECOM and Assistant Secretary of the Air Force for Acquisition (AQS).

[19]More money was given to DoD to acquire communications in such contingencies as Bosnia, according to our study sponsors.

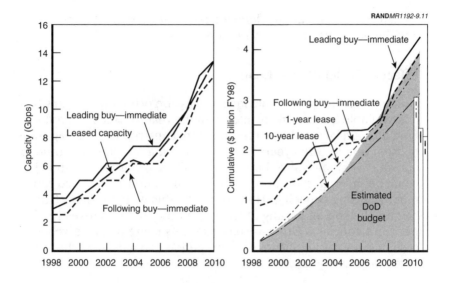

**Figure 9.11—Affordability of Strategies Within Estimated DoD Satellite Communications Budget**

tain the communications needed, we will assume that we cannot plan on additional money to fund communications in contingencies. Our assumed budget for future communications capacity therefore may be conservative.

As can be seen on the right-hand side of Figure 9.11, only the ten-year lease strategy fits within the estimated budget. This is unfortunate, because both the "leading" and "following" acquisition strategies may be cheaper. (We say may, because we have not yet factored in the effect of order-receipt lags.)

We might ask if DoD could sell surplus capacity on owned or leased assets to recoup a portion of their costs. Because they use military frequencies, DoD-unique satellites might not have a market outside of allied militaries. And these militaries might have surge demands for the same contingencies as those involving U.S. forces. Leases may be sublet, although the transaction costs might reduce the savings available. Answers to these questions require a much deeper look into regional military demand, commercial market transactions, and the practices of capacity brokers.

## Effect of Order-Receipt Lag

Unless a satellite is available for purchase on orbit, the year in which it is available for use will be later than the year in which it is ordered. For commercial satellites ordered before construction, this lag could be from one to three years for construction and additional time for launch.[20] For specialized military satellites, order-receipt time lags could be three, five, or more years.[21] In the interim, the need can be met by leasing capacity. Unfortunately, this would require DoD to pay for the cost of leasing at the same time that it is investing in a purchase contract. Order-receipt lags thus result in a relatively high cash outflow during the investment period and a delay of the antici-pated savings expected from an owned asset.[22]

Figure 9.12 compares the cumulative costs of leasing 1 Gbps of ca-pacity for one and ten years with the cumulative cost of purchasing that capacity with a time lag of zero, three, and five years.[23] For order-receipt lags greater than zero, the cost of leasing the capacity while waiting for receipt of the satellite is as shown. We can see from the figure that, with our baseline set of expectations about demand and price, purchases with no lag compare favorably with a ten-year lease.

However, if the order-receipt lag is three years, there is no difference in cumulative expected cost at year 13 between buying and leasing at the ten-year rate. That is, 13 years of leasing at the ten-year rate costs the same as leasing for three years at the one-year rate and then buying and operating the satellite. If the order-receipt lag is five

---

[20]These order-receipt lags are consistent with the INTELSAT 7-, 8-, and 9-series spacecraft.

[21]The 1999 President's Budget has three years of payments before the first Gapfiller is deployed.

[22]Because the savings are delayed, they are less valuable at present; in other words, they are discounted more heavily.

[23]The graph is based on the costs shown in Table 7.1. It assumes that the purchase price of the satellite is spread evenly over the period from the order year through the receipt year.

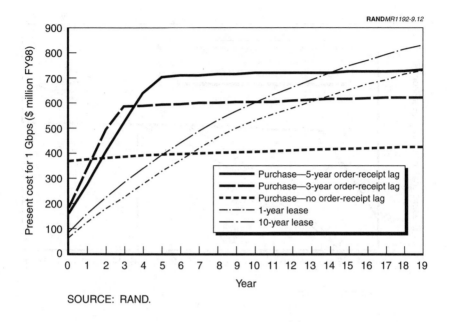

SOURCE: RAND.

**Figure 9.12—Present Cost of 1 Gbps as a Function of Strategy and Year**

years, DoD should be indifferent between buying the satellite and leasing at the one-year rate.

But what about the follow-on satellite? If DoD were to keep the constellation going indefinitely, then it should be able to order a new satellite before the first reaches expected end-of-life. DoD would not be paying development and construction money for the subsequent satellites at the same time it was leasing needed capacity. However, the important effect is that DoD would be paying for a satellite over some period before it received any benefits. We can visualize the comparison between buying a satellite with a three-year lag and leasing as shown in Figure 9.13.

In this example, DoD pays for the construction of a satellite in four increments—at the beginning of years one, two, and three, and when the satellite is delivered at the end of year three and begins operations. Alternatively, DoD could begin one- or ten-year leases at

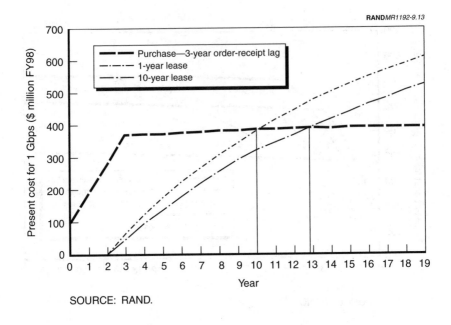

SOURCE: RAND.

**Figure 9.13—Present Cost of Follow-On Satellites**

the end of year three for the same amount of capacity. After seven years of satellite operations, the present value of the one-year lease cost finally exceeds the present value of purchasing the satellite. The satellite would need to operate for ten years before the cumulative cost of a ten-year lease would equal the cost of buying the satellite. These results are identical to those presented in Figure 9.12.

The effect of a three-year order-receipt lag on the capacity and cost streams from 1998 to 2010 are shown in Figure 9.14.

The net cost of the "leading" purchase strategy with a three-year order-receipt lag is roughly equal to that of the 10-year lease strategy. The cost of one- and ten-year lease strategies as a percentage of "leading" acquisition strategies with no lag, and order-receipt lags of three and five years, is shown in Table 9.1.

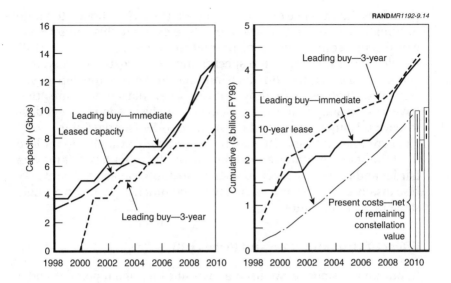

**Figure 9.14—Effect of a Three-Year Order-Receipt Lag on Expected Cost**

We can also use the results shown in Figure 9.13 to estimate the effect of variations in satellite lifetime on expected cost. If the satellite lasts longer than ten years, its expected cost relative to one- and ten-year leases declines. For example, after year 13 the purchase option, even with the three-year order-receipt lag, is once again cheaper than the ten-year lease option. Conversely, if the satellite fails prematurely, its expected cost rises.

**Table 9.1**

**Ratio of Lease Cost to Purchase Cost for 1 Gbps of Capacity**

| Lease | Lease/Purchase Cost Ratio— No Order- Receipt Lag | Lease/Purchase Cost Ratio— Three-Year Order- Receipt Lag | Lease/Purchase Cost Ratio— Five-Year Order- Receipt Lag |
|---|---|---|---|
| One-year | 1.52 | 1.16 | 1.04 |
| Ten-year | 1.26 | 0.96 | 0.86 |

How does the variance of expected life affect cost? Although too little information exists on military satellites to estimate this variance, we can discuss some aspects of premature failure. Generally, DoD should be risk neutral with respect to monetary costs, and therefore should make acquisition decisions based upon expected costs. However, satellites owned by DoD are often placed on routes that have less capacity than DoD anticipates will be needed to meet day-to-day and surge demand. If the satellite suffers partial failure, sufficient commercial capacity may be immediately available to replace it. If the satellite suffers a total failure, on the other hand, it may be harder and more expensive to obtain substitute capacity. There may be a disadvantage, therefore, in having too much capacity on a single satellite on a thin route.[24]

## Effect of Discount Rate and Price Drift

In our analysis above, we used a discount rate of 3.6 percent and a price trend of 4 percent per year. What if the discount rate was higher? What if prices declined more dramatically, or did not decline at all?

Higher discount rates result in greater reductions in the present value of future expenditures. Thus, strategies with a cost profile weighted more toward the future will benefit more from higher discount rates. Lease strategies have such a cost profile when compared with acquisition strategies. Thus, higher discount rates would decrease the cost "premium" for leasing strategies relative to acquisition. One-year leases are more sensitive to discount rates than are ten-year leases, and increasing discount rates will narrow the price gap between them.[25] Of course, the relations described here are reversed if discount rates are reduced.

A greater downward communications price drift would reduce the cost of future investments. It would reduce the price difference be-

---

[24]Of course, the same could be said if DoD leased most of its needed capacity in a thin market on a single commercial satellite. The key is to diversify the sources of communications.

[25]If, in fact, substantial sums are committed up front for the ten-year leases to provide for termination liability, then the ten-year leases will increase in price relative to one-year leases as the discount rate is increased.

tween a series of one-year leases and either a ten-year lease (assuming the annual payment rate is locked in at the current price) or acquisition of a satellite. (Again, the opposite is the case for a smaller price drift.)

Figure 9.15 shows the effect of the sum of the discount rate and annual rate of price change on the discounted cost of obtaining 1-Gbps of capacity. Again, the discounted cost of a series of one-year leases is favorably affected by both a higher discount rate and a higher rate of downward price drift. Thus, it falls more rapidly relative to the cost of acquisition than does the cost of a ten-year lease.

The price drift decreases the expected cost of satellite operations, satellite leases, and satellite purchases. For ten-year leases, however,

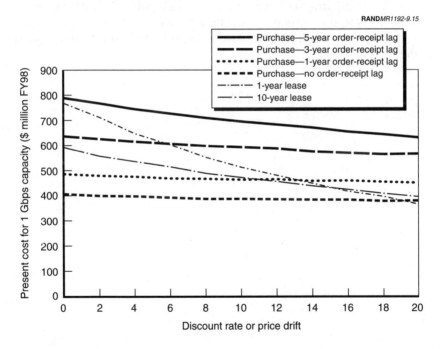

Figure 9.15—Effect of Discount Rate or Price Drift on Present Cost of 1 Gbps of Capacity as a Function of Strategy and Year

we assumed that the price was locked-in for the duration of the lease. Once the lease had lapsed, customers could then contract for new leases at reduced rates. For satellites with an order-receipt lag, we assumed that the purchase price was evenly spread over that lag. The purchase price declined for each of the future payments. For example, satellites bought with a five-year lag would be paid for in five payments from year zero to year four. The payment for years one through four would be decreased by the annual price decline rate.

## Effect of Demand Variations

All four strategies we examined are variations of "wait and see." That is, each year communications planners can reassess communications demand and market prices and decide whether to obtain more capacity and how much should be leased or bought. These strategies, therefore, each have inherent flexibility to handle variations in demand over time.

# Chapter Ten

# CONCLUSIONS

Based upon our analyses, we have reached the following general conclusions:

- DoD projects a large gap between its demand for communications and the capacity expected from present, planned, and programmed systems. Its options are to limit the amount of communications available to users, buy more DoD-unique systems, or employ commercial systems. Commercial leases provide a valuable option to increase capacity even when DoD buys unique systems.

- Cost is not the only criterion—sometimes DoD needs to pay more for military-unique operational capabilities. Where access, control, and protection are high priorities, DoD will have to rely on its own assets. Where these characteristics are not truly essential, DoD should be able to use commercial capacity while taking steps to improve access to, control over, and protection of these systems to the levels needed. Lease contracts should include the rights to switch transponders between beams as needed.

- DoD must develop operational concepts that maximize its flexibility in employing commercial and DoD systems in order to meet both day-to-day and contingency demand in a way that does not make it vulnerable to enemy disruption, technical failures, or market forces.

- It may be more economical to make long-term commitments and "waste" some capacity than to underestimate need and make up the shortfall with short-term service contracts. The degree to which this is true depends on the expected demand and the long-/short-term price ratio.

- A three-year lag between the order and receipt of a DoD-unique satellite reduces the premium for ten-year leases to zero, and a five-year lag results in near parity between purchasing a satellite and leasing the same capacity with one-year leases. Therefore, expected savings should not motivate buying DoD-unique satellites. DoD should make the choice based on the operational characteristics needed.

Although we anticipate exponential growth in commercial communications capacity during the next decade, this is not certain—and growth in supply may not be evenly distributed over the globe. Therefore, DoD should not assume that commercial capacity will be immediately available in every region all of the time. DoD reviews its communications needs annually, and this review includes projecting demand in each region. This review might also incorporate considerations of the commercial market in the following way:

- Compare projected demand with capacity supplied by DoD-owned systems and leased capacity.

- Determine if expected costs could be decreased by adding or relinquishing capacity (e.g., by using the methods described in Chapter 9).

- Determine contemporaneous prices for buying or leasing capacity.

- Buy or lease capacity as it decreases expected costs.

Regulatory changes, such as the reclassification of COMSAT as a nondominant carrier and the allowance of direct access for users, may significantly change prices offered for leased capacity. In general, these changes are expected to lead to price reductions. Even so, the published tariffs may describe less of the lease market in the future as more companies deploy their own systems or broker capacity provided by others. The DoD will need to develop new sources of information on the availability and prices of commercial capacity.

Currently, several DoD initiatives are underway to employ commercial communications. The largest on several dimensions is the Commercial Satellite Communications Initiative, or CSCI, which is managed by DISA. An in-depth assessment of these efforts might provide a better understanding of the current demand for communications and the ways in which commercial capabilities might be put to more effective and efficient use.

# STANDARD FREQUENCY DESIGNATIONS

| | Frequency |
|---|---|
| VLF | 3 kHz–30 kHz |
| LF | 30 kHz–300 kHz |
| MF | 300 kHz–3 MHz |
| HF | 3 MHz–30 MHz |
| VHF | 30 MHz–300 MHz |
| UHF | 300 MHz–3 GHz |
| Military UHF | 225 MHz–400 MHz |
| SHF | 3 GHz–30 GHz |
| EHF | 30 GHz–300 GHz |
| L | 1.0 GHz–2.0 GHz |
| S | 2.0 GHz–4.0 GHz |
| C | 4.0 GHz–8.0 GHz |
| X | 8.0 GHz–12.0 GHz |
| Ku | 12.0 GHz–18.0 GHz |
| K | 18.0 GHz–27.0 GHz |
| Ka | 27.0 GHz–40.0 GHz |
| V | 40.0 GHz–75 GHz |
| W | 75 GHz–110 GHz |

SOURCE: IEEE Standard Letter Designations for Radar Frequency Bands, IEEE Standard 521-1984, reaffirmed 1989.

# CURRENT DoD COMMUNICATIONS SATELLITES

| System | Frequency Band | User |
|---|---|---|
| Ultra-High Frequency Follow-On (UFO) | UHF (243–318 MHz) | Narrowband communications for ships, aircraft, submarines, and ground forces |
| | SHF (8-GHz uplink only) | Uplink only for fleet broadcast systems |
| | EHF (44-GHz uplink/ 20-GHz downlink) | Some low-data-rate protected channels |
| | | GBS package providing theater data broadcasts |
| Defense Satellite Communications System (DSCS) | SHF (8-GHz uplink/ 7-GHz downlink) | Fixed and deployed military users as well as government agencies |
| | | Some protected wideband capacity available |
| Military, Strategic, and Tactical Relay Satellite (MILSTAR) | EHF (44-GHz uplink/ 20-GHz downlink) | Fixed and deployed military and government users needing the highest available levels of secure, survivable, and protected communications |

# CURRENT PERFORMANCE COST ESTIMATES

# GRAPHICAL METHOD FOR COMMUNICATIONS PLANNING UNDER UNCERTAINTY

This appendix shows the derivation of the graphical method of find-ing the optimal amount of communications capacity discussed in Chapter Nine.

A communications planner faces the basic problem of an uncertain demand for satellite communications capacity. He can buy some fixed capacity before the actual demand can be determined. He can buy additional capacity on the spot market when the actual demand is known. How much fixed capacity should he buy to minimize ex-pected cost?

## A SIMPLE MATHEMATICAL MODEL

Suppose that demand for satellite communications x is distributed as $f(x)$. A communications planner can buy fixed capacity a for total cost $p_a a$ before the actual demand is known. He can buy residual capacity needed on the spot market for $p_s(x-a)$, if x is greater than a. What is the value of a that minimizes expected cost?

We can write the following expression for the expected cost C:

$$C = p_a a + p_s \int_a^\infty (x-a)f(x)dx \qquad (1)$$

where $p_a$  is the price of a,

  a    is the amount of fixed capacity,

  $p_s$  is the spot market price,

  x    is communications demand, and

  f(x)  is the distribution of demand x.

The first term in the expression, $p_a a$, gives the total cost of fixed capacity a. The second term in the expression,

$$p_s \int_a^\infty (x-a)f(x)dx,$$

gives the expected expenditure on spot market communications capacity, given that fixed capacity is a.

We can find the value of a that minimizes the expected cost C by using the first-order condition to solve for the optimal value a*. The first-order condition is simply that the first derivative of C with respect to a is equal to zero. Equation (2) gives the first-order condition.

$$\frac{\partial C}{\partial a} = p_a - p_s \int_a^\infty f(x)dx = 0 \tag{2}$$

We can rewrite Eq. (2) in terms of the cumulative density function,

$$\frac{\partial C}{\partial a} = p_a - p_s \left[1 - F(a)\right] = 0 \tag{3}$$

where $F(a) = \int_{-\infty}^a f(x)dx$. Solving Eq. (3) for a gives

$$a^* = F^{(-1)}\left(1 - \frac{p_a}{p_s}\right) \tag{4}$$

where $F^{(-1)}(\cdot)$ denotes the inverse of $F(\cdot)$ (i.e., $F^{(-1)}[F(x)] = x$). $a^*$ is the amount of fixed capacity that minimizes the expected cost.

We can verify that we are at a local minimum by examining the second-order condition:

$$\frac{\partial^2 C}{\partial a^2} = p_s f(a) > 0 \tag{5}$$

which indicates $a^*$ is a local minimum.

## GRAPHICAL INTERPRETATION OF THE SOLUTION

Figure C.1 gives a graphical interpretation of the solution, showing an illustrative cumulative distribution function that is read with respect to the left-hand axis. The price ratio is read off of the right-hand axis, which is numbered so that if x is the corresponding number on the left-hand axis, 1 – x is the number on the right-hand axis (e.g., if 0.1 is the number on the left-hand axis, 0.9 is the corresponding number on the right-hand axis). The graphical process of drawing a line from the price ratio on the right-hand axis to the cumulative distribution function line and then dropping a line to the horizontal axis to find the optimal capacity is equivalent to solving the equation below to find the optimal capacity.

$$a^* = F^{(-1)}\left(1 - \frac{p_a}{p_s}\right)$$

## KEY ASSUMPTIONS

This model, while giving a general solution, is based on several key assumptions:

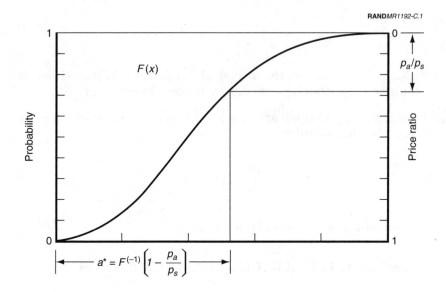

**Figure C.1—Graphical Interpretation of Optimal Solution**

- Minimization of present value of expected cost is the appropriate objective
- Demand is exogenous
- Relative prices are exogenous
- Prices are linear in quantity.

Let's briefly discuss each of these assumptions.

*Minimization of present value of expected cost is the appropriate objective.* This assumption means that the objective function considers only expected cost and not the possible variance in cost. If the decisionmaker is risk averse, this would not be an appropriate assumption. Because the decisionmaker in this case is acting in the interests of the U.S. government, and the government is highly diversified, it is reasonable to assume that the government is risk neutral, i.e., that the government cares only about expected cost and not about possible variance in cost.

*Demand is exogenous.* This assumption means that the demand given by the distribution function is not a function of the price of communications capacity. Demand in this case is set externally and is not a decision variable; the only decision variable available to the decisionmaker is the amount of fixed capacity to buy. The decisionmaker has no impact on the amount demanded.

*Relative prices are exogenous.* This assumption means that the decisionmaker is a "price-taker" and decisions made by the decisionmaker do not influence the market price.

*Prices are linear in quantity.* This assumption means that the total cost of a particular quantity of communications capacity can be expressed as the product of the quantity and a price. That is, there are no "quantity breaks."

It is unlikely that analytic solutions, much less graphical solutions, exist for models that violate these assumptions.

ACM—see Association for Computing Machinery.

Association for Computing Machinery, Committee on Computers and Public Privacy, "Issue 81 Forum on Risks to the Public in Computers and Related Systems," Peter G. Neumann, moderator, in *RISKS-FORUM Digest*, Vol. 6, May 9, 1988.

Biddle F., J. Lippman, and S. Mehta, "Satellite Failure Severs Lifeline That Few Knew Even Existed," *Wall Street Journal*, May 21, 1998.

Blake, Linda, and Jim Lande, *Trends in the U.S. International Telecommunications Industry*, Federal Communications Commission, Washington, D.C., August 1998.

Briefing by Cdr Baccioco to the SWARF, August 1997, Peterson AFB, Colorado, and early Capstone analyses.

Briefings from the 1997 MILSATCOM Senior Warfighter Forum (SWARF).

Clinton, William J., *A National Security Strategy for a New Century*, The White House, Washington, D.C., October 1998.

COMSAT Corporation, *COMSAT World Systems Tariff*, FCC No. 3, November 14, 1998.

Department of Defense, *Program Budget Decision 417C*, Pentagon, Washington, D.C., December 11, 1998.

Department of Defense, *Spectrum Management Issues Associated with MILSATCOM Architectures*, prepared for DoD Office of Space

Architect (DoDOSA), Joint Spectrum Center, Annapolis, Maryland, September 1996.

Department of the Air Force, *Fiscal Year 2000/2001 Budget Estimates Procurement Program, Other Procurement*, February 1999.

DoD—see Department of Defense.

Estes, General Howell M. III, in *Advanced Military Satellite Communications Capstone Requirements Document,* HQ SPACE-COM, Peterson Air Force Base, Colorado, April 24, 1998.

Euroconsult, *World Satellite Communications and Broadcasting Markets Survey:  Prospects to 2006,* WSMS #2, Paris, France, Summer 1996.

FCC—see Federal Communications Commission.

Federal Communications Commission, *Circuit Status Data,* Report IN 99-4, Section 43.82, 1997.

Federal Communications Commission, Order 97-121, April 4, 1997.

Federal Communications Commission, Order 99-17, August 1998.

Federal Communications Commission, Order 99-236, September 1999.

General Accounting Office, *Defense Satellite Communications: Alternative to DoD's Satellite Replacement Plan Would Be Less Costly,* GAO/NSIAD-97-159, Washington, D.C., July 1997.

Headquarters United States Space Command, *Advanced Military Satellite Communications Capstone Requirements Document,* Peterson Air Force Base, Colorado, April 24, 1998.

Headquarters United States Space Command, "Joint Wideband Gapfiller Concept of Operations" (draft), Peterson Air Force Base, Colorado, December 1, 1998.

HQ USSPACECOM—see Headquarters United States Space Command.

INTELSAT, *Annual Report 1998,* Washington, D.C.

INTELSAT, *Earth Station Service Capabilities*, Washington, D.C., May 1996.

International Telecommunications Union, *Making Space in Space*, 1997.

Joint Chiefs of Staff, *The National Military Strategy*, www.jcs.mil.

Martin, Donald H., *Communications Satellites 1958–1995*, The Aerospace Corporation, El Segundo, CA, ISBN 1-884989-02-0, 1996.

Merrill Lynch, Pierce, Fenner & Smith Inc., *Global Satellite Marketplace 98*, 1998.

OMB (Office of Management and Buget), Circular A-94, November 1992.

Raduege, Major General Harry D., Jr., *Operation Desert Thunder, New Dimensions for C4I in Expeditionary Warfare*, USAF CENTCOM J-6, 1998.

Shalikashvili, General John M., Chairman of the Joint Chiefs of Staff, *Joint Vision 2010*, 5126 Joint Staff, Pentagon, Washington, D.C., 1997a.

Shalikashvili, General John M., Chairman of the Joint Chiefs of Staff, *Shape, Respond, Prepare Now: A Military Strategy for a New Era*, 5126 Joint Staff, Pentagon, Washington, D.C., 1997b.